Single-Port Robotic Surgery in Urology

The New Beginning After the Advent of Dedicated Platforms

Single-Port Robotic Surgery in Urology

The New Beginning After the Advent of Dedicated Platforms

Edited by

Riccardo Bertolo
Department of Urology, "San Carlo di Nancy" Hospital, Rome, Italy

Jihad Kaouk
Department of Urology, Glickman Urological and Kidney Institute, Cleveland Clinic, Cleveland, OH, United States

ELSEVIER

ACADEMIC PRESS
An imprint of Elsevier

Academic Press is an imprint of Elsevier
125 London Wall, London EC2Y 5AS, United Kingdom
525 B Street, Suite 1650, San Diego, CA 92101, United States
50 Hampshire Street, 5th Floor, Cambridge, MA 02139, United States
The Boulevard, Langford Lane, Kidlington, Oxford OX5 1GB, United Kingdom

ISBN: 978-0-323-91906-7

For Information on all Academic Press publications
visit our website at https://www.elsevier.com/books-and-journals

Publisher: Stacy Masucci
Editorial Project Manager: Timothy Bennett
Production Project Manager: Maria Bernard
Cover Designer: Christian J. Bilbow

Typeset by MPS Limited, Chennai, India

Contents

List of contributors

Ronney Abaza
Central Ohio Urology Group, Mount Carmel St. Ann's Hospital, Westerville, OH, United States

Mahmoud Abou Zeinab
Department of Urology, Glickman Urological and Kidney Institute, Cleveland Clinic, Cleveland, OH, United States

Mutahar Ahmed
Department of Urology, Hackensack University Medical Center, Hackensack, NJ, United States

Mohamad E. Allaf
The James Buchanan Brady Urological Institute, Johns Hopkins University School of Medicine, Baltimore, MD, United States

Alireza Aminsharifi
Department of Urology, Hershey Medical Center, Pennsylvania State University, Hershey, PA, United States

Riccardo Autorino
Division of Urology, Department of Surgery, VCU Health, Richmond, VA, United States

Christopher Bednarz
Division of Urology, Department of Surgery, VCU Health, Richmond, VA, United States

Riccardo Bertolo
Department of Urology, San Carlo di Nancy Hospital, Rome, Italy

Mubashir Billah
Department of Urology, Hackensack University Medical Center, Hackensack, NJ, United States

Grace Chen
Department of Urology, University of Illinois at Chicago, Chicago, IL, United States

Nathan Cheng
Department of Urology, Hackensack Meridian School of Medicine, Nutley, NJ, United States

Marcio Covas Moschovas
AdventHealth Global Robotics Institute, Urology, Celebration, FL, United States; University of Central Florida (UCF), Orlando, FL, United States

Simone Crivellaro
Department of Urology, University of Illinois at Chicago, Chicago, IL, United States

Prokar Dasgupta
King's Health Partners, London, United Kingdom

Andrew T. Gabrielson
The James Buchanan Brady Urological Institute, Johns Hopkins University School of Medicine, Baltimore, MD, United States

Juan Garisto
Division of Urology, Veterans Affairs Medical Center, Boston, MA, United States; Harvard Medical School, Boston, MA, United States; Department of Surgery, Brigham and Women's Hospital, Harvard Medical School, Boston, MA, United States

Christian Huge
Department of Urology, Hershey Medical Center, Pennsylvania State University, Hershey, PA, United States

Jihad Kaouk
Department of Urology, Glickman Urological and Kidney Institute, Cleveland Clinic, Cleveland, OH, United States

Paul Kogan
Division of Urology, Department of Surgery, Washington University in St. Louis, St. Louis, MO, United States

Michelle Kruse
Department of Urology, Hackensack University Medical Center, Hackensack, NJ, United States

Vipul Patel
AdventHealth Global Robotics Institute, Urology, Celebration, FL, United States; University of Central Florida (UCF), Orlando, FL, United States

Phillip M. Pierorazio
Division of Urology, Perelman School of Medicine at the University of Pennsylvania, Philadelphia, PA, United States

Venkat M. Ramakrishnan
Department of Surgery, Brigham and Women's Hospital, Harvard Medical School, Boston, MA, United States; Division of Urology, Brigham and Women's Hospital, Boston, MA, United States

Fairleigh Reeves
Guy's and St Thomas' Hospital, London, United Kingdom

Fahad Sheckley
Department of Urology, Hackensack University Medical Center, Hackensack, NJ, United States

Michael Stifelman
Department of Urology, Hackensack Meridian School of Medicine, Nutley, NJ, United States

Shirley L. Wang
The James Buchanan Brady Urological Institute, Johns Hopkins University School of Medicine, Baltimore, MD, United States

Wesley M. White
Division of Urologic Surgery, The University of Tennessee Medical Center, Knoxville, TN, United States

Preface

Urology is a technology-driven specialty and has traditionally integrated new minimally invasive approaches at its early stage.

In the last decade, while looking for the least invasive approach, an attempt to shift from the traditional "multiincision" laparoscopic surgery to the "single-incision" laparoscopic surgery (the so-called "laparoendoscopic single-site surgery"—LESS) was observed. However, several drawbacks were observed with the laparoscopic single-site approach, including limited surgical movements and external clashing of instruments. These made the laparoscopic approach challenging to be performed, with a long learning curve even for skilled laparoscopic surgeons, resulting in limited diffusion of pure LESS.

The introduction of dedicated purpose-built devices and instruments for robotic platforms renewed the interest for this approach, paving the way to robotic LESS. Robotic LESS was seeking an easing of the known drawbacks typical of the pure LESS approach.

The first series of robotic LESS in urology was described in 2008 by Jihad Kaouk at Cleveland Clinic.

At that time, Riccardo Bertolo was just a medical student in Turin, getting in contact with the "wonders" of robotic urology inspired by Prof. Francesco Porpiglia.

Since then, several studies on robotic LESS applied to many urological procedures followed. Despite some reported benefits, consistent drawbacks including the external clashing and the limited accessibility for the assistant surgeon remained, so that robotic LESS did not spread.

Riccardo's mentor, Prof. Porpiglia, had the great idea to invent a unique meeting, aimed to show the most recent technological innovations introduced in the daily armory of urologic surgeons, and to determine if they could have ever replaced the current approaches in the treatment of urologic diseases.

It was in 2013 during the first edition of this Techno-Urology Meeting in Turin that Jihad and Riccardo met each other for the first time.

Jihad was the invited cochairman of the meeting and performed a robot-assisted LESS partial nephrectomy. Not surprisingly, Riccardo was fascinated by this new technology and technique, in hands so capable of making the most of it.

Collaborations started after that meeting, to the point that, in 2017, Riccardo moved as a urologist and PhD student in bioengineering to Cleveland Clinic to work with Jihad for a 1-year fellowship in robotic urology.

That was an unmissable chance for him to surf the promising wave of the new robotic platform coming to Cleveland Clinic, purpose-built to perform robotic LESS surgery, potentially representing a step forward in robotic urology.

Led by Jihad Kaouk's vision and inspiration, preclinical and preliminary clinical experiences were performed in all the major urological surgical interventions by using this novel platform, with Riccardo and Jihad, together with another energetic fellow, Dr. Juan Garisto, publishing around 50 scientific papers together as a team, during that 1-year time span.

The investigated features of the novel robotic platform were democratizing robotic LESS, so that it is being nowadays performed by many surgeons across the United States and far East, thus coming out of the niche.

After more than 3 years, beyond their working collaboration, a friendship has formed between Riccardo and Jihad. With this novel platform being on the market, they felt to capitalize it and to invest their effort and personal experience to offer other robotic teams around the world an educational support in the field.

They believed a comprehensive book would have been welcome. Thus they have invited the top expert robotic single-port teams from the United States, who contributed to this work, trying to highlight the new standards of single-port robotic urology today, while showing its future directions.

We are delighted that this book has come to a fruitful conclusion after a year of hard work.

Our thanks go to Stacy Masucci and Timothy Bennett from the Elsevier team for their great editorial assistance.

We are grateful to our families for their support and tolerance of our high professional workload.

We thank everybody for their motivation as well as for their criticism in the past—all of you have brought forward the vision of minimizing the invasiveness of minimally invasive urology, while keeping the highest standards.

A dream is becoming reality...what was just a pioneer experience some years ago is now taking place as a true indication in the world of robotic urology!

Riccardo Bertolo and Jihad Kaouk together in Dr. Kaouk's office at the Glickman Urological and Kidney Institute, Cleveland Clinic, Cleveland, OH, USA, in 2018.

Riccardo Bertolo[1] and Jihad Kaouk[2]

[1]*Department of Urology, San Carlo di Nancy Hospital, Rome, Italy*

[2]*Department of Urology, Glickman Urological and Kidney Institute, Cleveland Clinic, Cleveland, OH, United States*

CHAPTER

The evolution of robotic single-port dedicated platforms

1

Riccardo Autorino[1], Christopher Bednarz[1], Fairleigh Reeves[2] and Prokar Dasgupta[3]

[1]*Division of Urology, Department of Surgery, VCU Health, Richmond, VA, United States*

[2]*Guy's and St Thomas' Hospital, London, United Kingdom*

[3]*King's Health Partners, London, United Kingdom*

1.1 Introduction

While the idea of using a "single port" for diagnostic and simple interventions has been in place in the surgical field for many decades,[1] it was only in the early 1990s that the first complete extirpative procedure was reported in gynecology.[2] In the mid-2000s, the fascinating concept of "scarless surgery" was revived creating momentum for a new "wave" of preclinical and clinical investigations in different surgical specialties. Both "natural orifice transluminal endoscopic surgery" and "laparoendoscopic single-site surgery" (LESS) were developed to further reduce the morbidity and scarring associated with minimally invasive surgical intervention. These techniques shared a common underlying "hypothesis," that a reduction in the number of transcutaneous points of access (i.e., trocar sites) may ultimately translate into better outcomes, less complications, and faster postoperative recovery.[3,4]

In history of LESS in urology started in 2005 with a report by Hirano et al. on a retroperitoneoscopic adrenalectomy via a "single port."[5] In 2007 Rane et al. presented at the 25th World Congress of Endourology in Cancun the first successful LESS case in urology.[6] Soon after, Cadeddu et al. from UTSW Dallas independently reported the first series of LESS transumbilical nephrectomies.[7] Kaouk and Gill from the Cleveland Clinic also pioneered the early adoption of this groundbreaking concept by showing the "feasibility" of the entire spectrum of urologic procedures.[8–10] The initial enthusiasm was, however, tempered by the technical challenges related to the use of a single access point, forcing the surgeon to work with laparoscopic instruments in parallel and causing by default reduced range of instrument movements, limited extraabdominal working space, unfavorable ergonomics, and more importantly, increasing the risk of the procedure. These limitations de facto restrained the implementation of LESS to few surgeons worldwide and in few selected cases.[11]

Single-Port Robotic Surgery in Urology. DOI: https://doi.org/10.1016/B978-0-323-91906-7.00009-0

The recognized benefits of the robotic-assisted laparoscopy over conventional laparoscopy include superior ergonomics, optical magnification of the operative field, enhanced surgeon dexterity, and precision of surgical manipulation. It was, therefore, natural to try to use the da Vinci robotic system to facilitate LESS. In 2009 Kaouk and his team reported the first successful series of "single-port" robotic procedures in humans, including cases of radical prostatectomy, dismembered pyeloplasty, and radical nephrectomy. The authors immediately noted less challenging intracorporeal dissection and suturing using robotic instrument compared to standard LESS.[12] With the adoption of robotic technology, a new era seemed to start, but it soon became clear that technical challenges still existed, given the lack of a purpose-built robotic system.[13]

In this chapter, we discuss the evolution of robotic "single-site" or "single-port" surgery in urology, including trends, techniques, challenges, outcomes, and technological advances made until the release of the first commercially available purpose-built robotic platform (da Vinci SP System, Intuitive Surgical, Sunnyvale, CA, USA).

1.2 Da Vinci Single Site (VESPA)

Given the limitations of using the standard multiport da Vinci Si robot, Intuitive Surgical initially developed and launched a novel set of single-site instruments and accessories specifically dedicated to LESS (Fig. 1.1).[14] The set included a

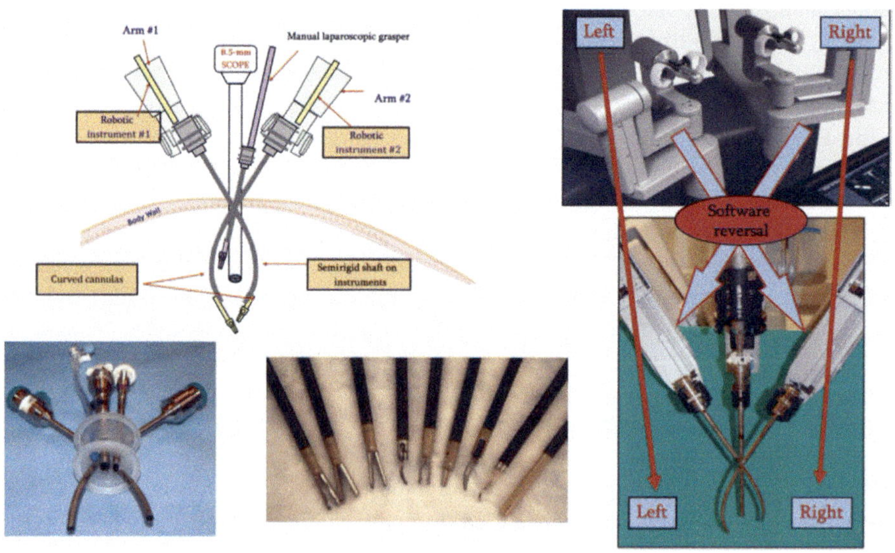

FIGURE 1.1

Da Vinci "Single-Site" platform (VESPA) (Intuitive Surgical).

multichannel access port with room for four cannulas and an insufflation valve: two curved cannulas were for robotically controlled instruments, and the other two cannulas were straight, one 8.5 mm cannula for the robotic endoscope and a 5-mm for bedside-assistant port. The curved cannulas were integral to the system, since their configuration allowed the instruments to be positioned to achieve triangulation of the target anatomy. This triangulation was achieved by crossing the curved cannulas midway through the access port. Same-sided hand–eye control of the instruments was maintained through assignment of the Si system software enabling the surgeon's right hand to control the screen right instrument even though the instrument was in the left robotic arm and, reciprocally, the left hand to control the screen left instrument which was in fact the right robotic arm. Robotic arm collisions were minimized externally because the curved cannulas angle the robotic arms away from each other. The single-site instruments and accessories were intended to be used with the da Vinci Si surgical system, being of similar construction to existing EndoWrist instruments, except they did not have a wrist at the distal end of the instrument, which turned out of a major limitation to future clinical adoption.

1.2.1 Preclinical experience

This first robotic instrumentation purposely made for LESS procedures was initially tested in preclinical animal model in 2010 by Haber et al. at the Cleveland Cinic.[15] Sixteen kidney procedures were performed without additional ports or need for conversion. During this feasibility evaluation, limitations of the platform were noted, including the lack of articulation at the tip of the instruments, making intracorporeal suturing more challenging. Soon after, Kaouk et al. reported the use of the "second generation" of da Vinci Single-Site instruments for R-LESS to perform different kidney procedures in the cadaver model.[16] The procedures were completed successfully without the addition of extra ports. The working space was not problematic for the surgeon, but significant collision against the robotic arms was experienced by the assistant, which at times restricted retraction and suction. A positive feature was the easy insertion and exchange of the instruments. The lack of wrist articulation was confirmed as the main limitation of these instruments, particularly for procedures involving suturing.

1.2.2 Clinical experience

In the clinical arena, cholecystectomy has been the most widely used application of LESS within general surgery. This has also been the case for the "single-site" da Vinci platform, which was launched by Intuitive to primarily "target" general surgeons performing cholecystectomies. In 2011 two groups independently reported the first clinical experience with the novel da Vinci Single-Site platform.[17,18]

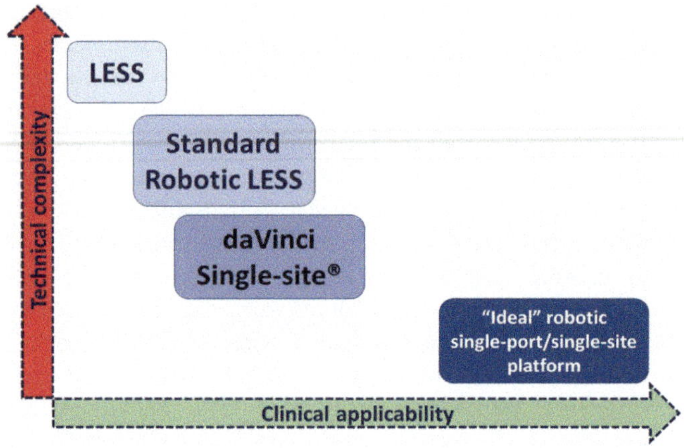

FIGURE 1.2

Transitioning from LESS to robotic single-port surgery.

In urology, Cestari et al. from Italy tested the technical feasibility and short-term perioperative outcomes of the da Vinci Single-Site instrumentation platform for the treatment of upper ureteropelvic junction obstruction in a selected group of patients.[19] Another study from Italy confirmed single-site robotic pyeloplasty to be feasible in selected patients, with good cosmetic results and excellent short-term clinical outcomes.[20] Not many other clinical series in urology followed.[21]

The introduction of the single-site instrumentation was initially seen as a step forward, as it was addressing some of the existing drawbacks of robotic LESS, mainly the clashing and lack of triangulation. However, the lack of EndoWrist technology at the instrument tips, which certainly represent a key feature of robotic surgery, represented a main limitation and there the interest in urology for this platform quickly faded away. The "ideal" robotic platform for LESS should have a low external profile, the possibility of being deployed through a single access site, and the possibility of restoring intraabdominal triangulation while maintaining the maximum degree of freedom for precise maneuvers and strength for reliable traction (Fig. 1.2).

1.3 Da Vinci SP

1.3.1 SP999 prototype

In 2014 a group of investigators from the Cleveland Clinic and University of Lille reported a clinical study supported by the Intuitive Surgical on the use of the first ever robotic system specifically built for "single-port" robotic surgery, the da Vinci SP Surgical System (SP999).[22] As for the standard multiport da Vinci robot, the three major system components were: surgeon console, patient side cart, and vision cart (Table 1.1). The da Vinci SP EndoWrist (also known as EndoWrist SP) instruments had 7 degrees of freedom at the distal end, and the system software was designed to provide exact translation of the surgeon's hand and finger movements at the console to precise and tremor-free movements of the instruments. The EndoWrist SP instruments incorporated an additional joint providing an "elbow" so that instruments could triangulate at the surgical site through a single port. The EndoWrist SP instruments were also longer and incorporating a "snake-style" wrist like the standard 5-mm multiport EndoWrist instruments (Fig. 1.3). This seminal work paved the way for subsequent developments.

Table 1.1 Components of the da Vinci SP system (SP999) prototype.

Surgeon console (SP999)	• The operative surgeon sits at the surgeon console, views the surgical site in a 3D (stereo) viewer, and controls movements of the surgical instruments using two master controllers. • The surgeon console also includes a touch screen display for setting up preferences and operating parameters, and a foot switch that allows the surgeon to toggle between operating modes and activation of electrosurgical energy to instruments. • The surgeon console design on the da Vinci SP Surgical System is identical to the standard da Vinci system except for an extra foot switch to allow control of the instrument arm.
Patient side cart (PS999)	• The side cart incorporates four robotic manipulators moving the endoscopic instruments and camera. These manipulators are known as instrument drives (also called patient side manipulators) and serve as the slave of the master-slave teleoperation control loop. These instrument drives are mounted to an instrument arm attached to the cart through a series of setup joints. • The port clutch button is thus used to move the location of the instrument arm port relative to the patient. In addition to the port clutch button, an instrument arm clutch button releases the outer pitch and outer yaw axes of the instrument arm, allowing for the port to be pivoted or reoriented in the patient about its remote center without changing the port location. • The patient side cart of the da Vinci SP Surgical System SP999 uses the same base and column of the da Vinci Surgical System IS3000 patient side cart, but the configuration of the surgical arms and manipulators is changed to manipulate the unique SP999 instruments and camera through a single port.
Vision cart (VS999)	• Similar to da Vinci vision cart VS3000, except that the camera controller is uniquely designed to interface with a new camera module that is unique to the da Vinci SP system.

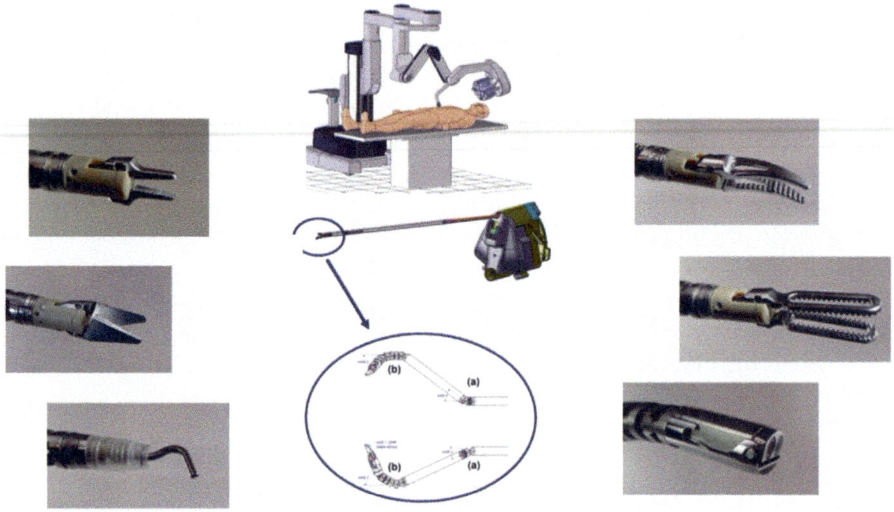

FIGURE 1.3

Details of the da Vinci SP EndoWrist instruments: they have 7 degrees of freedom at the distal end, incorporate an additional joint providing an "elbow" so that the instruments can triangulate at the surgical site through a single port, they are longer and feature a "snake-style" wrist like the 5-mm multiport EndoWrist instruments.

1.3.2 SP9018 system

More recently, an evolution of the initial SP robotic platform (known as "SP1098") was first investigated in preclinical setting,[23–26] and it finally received clearance by the US Food and Drug Administration (FDA) on May 31, 2018. Kaouk et al. reported the first case series of major urologic surgeries performed in living humans using this platform,[27] featuring three multijointed wristed instruments and a fully wristed three-dimensional HD camera, all emerging through a single multichannel 2.5 cm port, and the 360-degreee anatomic access allowed by a single robotic arm. The authors pointed out that some limitations are not completely solved with the SP robotic system because access to the surgical field for the assistant remains challenging. Moreover, an initial learning curve is expected as some differences can be perceived, particularly during intracorporeal suturing. Lenfant et al. recently published on the learning curve of SP robot-assisted extraperitoneal prostatectomy using the CUSUM method. Their analysis revealed that the learning curve based on operative time can be divided into four phases of approximately 30–40 cases. Similarly, in the first 50 cases, more complications were seen than expected. Beyond 120 cases, a mastery phase began, where the study-defined trifecta outcome criteria threshold was reached (surgical margin status, postoperative complications, and biochemical recurrence). The authors cautioned that the expertise of the study surgeon, as well as active preparation using simulation to facilitate the transition, may limit generalizability to the novice surgeon.[28]

1.4 Upcoming robotic platforms

While the current state of robotic surgery is largely dominated by the da Vinci surgical systems, new surgical robot might enter the clinical arena in the next 5 years, and additional purpose-built "single-port" systems might finally become available.[29–31]

Toronto-based Titan Medical has been in development with a single-port robotics system since at least 2016. The SPORT Surgical System features 3D-HD visualization, an ergonomic open workstation equipped with a 32″ 3D high-definition flat-screen monitor, a single-arm mobile patient cart with a multiarticulating endoscope providing 6 degrees of freedom and multiarticulated instruments with single-use replaceable tips. The surgeon console and the patient cart are the two main components of the system (Fig. 1.4). The surgeon is seated in an ergonomic position at the console, which can be individually adjusted, and operates the instruments using a natural handle interface of fingertip with clutching akin to open surgery. Several early preclinical animal-based studies have demonstrated the technical feasibility of the SPORT surgical system for general surgery and urologic surgery procedures.[31] In June 2020, Titan Medical developed an agreement with Medtronic to provide access to intellectual property and license to the single-port SPORT Surgical System, with the stated goal that each company will continue to develop their own robot.

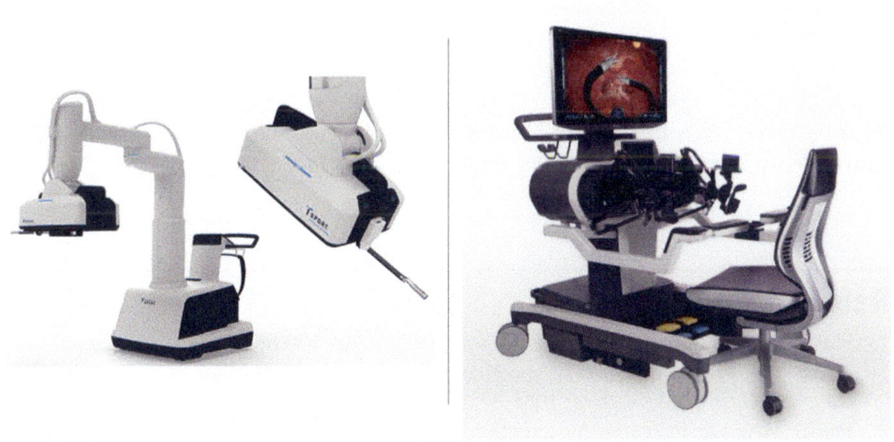

FIGURE 1.4

SPORT robotic system (Titan Medical Inc.).

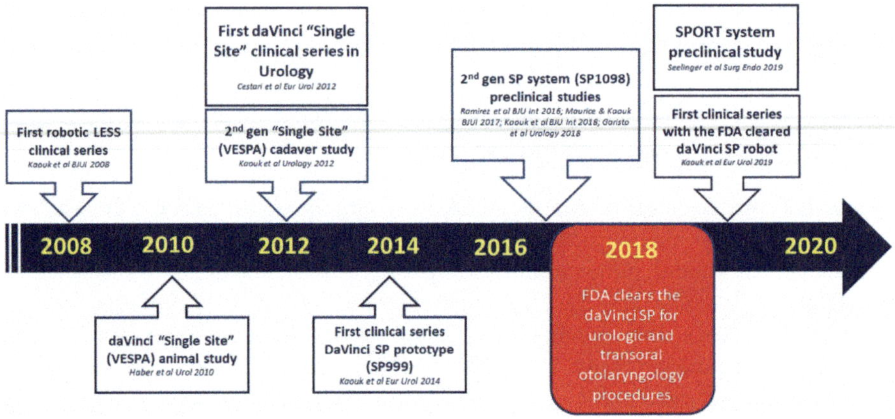

FIGURE 1.5

Timeline of the evolution of robotic single-port platforms: landmark studies.

Several other robotic prototypes for single-port surgery are being tested, and they remain at very early stages of development.[32]

1.5 Conclusion

After 10 years of evolution, robotic "single-site" or "single-port" surgery left its stage of infancy, finally entering the clinical arena (Fig. 1.5). The FDA approval and subsequent introduction in the market of the da Vinci SP system is allowing several centers in the United States (and soon in other parts of the world) to start structured robotic "single-port" programs. Since this is the first generation of this commercially available platform, it is expected that the future versions will further facilitate "single-port" surgery and expand its indications. Moreover, upcoming platforms from competing companies might enter the market and hopefully stimulate further innovation in the field.

References

1. Platteborse R. Laparoscopy, laparophotography, punch biopsy of the liver, gallbladder punch biopsy and collection of specimens of the peritoneal organs through a single trocar. *Acta Gastroenterol Belg*. 1961;24:696–700.
2. Pelosi MA. Laparoscopic hysterectomy with bilateral salpingo-oophorectomy using a single umbilical puncture. *N J Med*. 1991;88:721–726.
3. Gill IS, Advincula AP, Aron M, et al. Consensus statement of the consortium for laparoendoscopic single-site surgery. *Surg Endosc*. 2010;24:762–768.

4. Autorino R, Cadeddu JA, Desai MM, et al. Laparoendoscopic single-site and natural orifice transluminal endoscopic surgery in urology: a critical analysis of the literature. *Eur Urol*. 2011;59(1):26–45.
5. Hirano D, Minei S, Yamaguchi K, et al. Retroperitoneoscopic adrenalectomy for adrenal tumors via a single large port. *J Endourol*. 2005;19:788–792.
6. Rane A, Kommu S, Eddy B, Bonadio F, Rao P, Rao P. Clinical evaluation of a novel laparoscopic port (R-Port) and evolution of the single laparoscopic port procedure (SLiPP). *J Endourol*. 2007;21:A22–A23.
7. Raman JD, Bensalah K, Bagrodia A, Stern JM, Cadeddu JA. Laboratory and clinical development of single keyhole umbilical nephrectomy. *Urology*. 2007;70:1039–1042.
8. Kaouk JH, Haber G-P, Goel RK, et al. Single-port laparoscopic surgery in urology: initial experience. *Urology*. 2008;71:3–6.
9. White WM, Haber GP, Goel RK, Crouzet S, Stein RJ, Kaouk JH. Single-port urological surgery: single-center experience with the first 100 cases. *Urology*. 2009;74(4):801–804.
10. Desai MM, Berger AK, Brandina R, et al. Laparoendoscopic single-site surgery: initial hundred patients. *Urology*. 2009;74(4):805–812.
11. Kaouk JH, Autorino R, Kim FJ, et al. Laparoendoscopic single-site surgery in urology: worldwide multi-institutional analysis of 1076 cases. *Eur Urol*. 2011;60(5):998–1005.
12. Kaouk JH, Goel RK, Haber GP, Crouzet S, Stein RJ. Robotic single-port transumbilical surgery in humans: initial report. *BJU Int*. 2008;103:366–369.
13. Sorokin I, Canvasser NE, Irwin B, et al. The decline of laparoendoscopic single-site surgery: a survey of the endourological society to identify shortcomings and guidance for future directions. *J Endourol*. 2017;31(10):1049–1055.
14. Autorino R, Kaouk JH, Stolzenburg JU, et al. Current status and future directions of robotic single-site surgery: a systematic review. *Eur Urol*. 2013;63(2):266–280.
15. Haber GP, White MA, Autorino R, et al. Novel robotic da Vinci instruments for laparoendoscopic single-site surgery. *Urology*. 2010;76(6):1279–1282.
16. Kaouk JH, Autorino R, Laydner H, et al. Robotic single-site kidney surgery: evaluation of second-generation instruments in a cadaver model. *Urology*. 2012;79(5):975–979.
17. Kroh M, El-Hayek K, Rosenblatt S, et al. First human surgery with a novel single-port robotic system: cholecystectomy using the da Vinci single-site platform. *Surg Endosc*. 2011;25:3566–3573.
18. Wren SM, Curet MJ. Single-port robotic cholecystectomy: results from a first human use clinical study of the new da Vinci single-site surgical platform. *Arch Surg*. 2011;146:1122–1127.
19. Cestari A, Buffi NM, Lista G, et al. Feasibility and preliminary clinical outcomes of robotic laparoendoscopic single-site (R-LESS) pyeloplasty using a new single-port platform. *Eur Urol*. 2012;62:175–179.
20. Mathieu R, Verhoest G, Vincendeau S, Manunta A, Bensalah K. Robotic-assisted laparoendoscopic single-site radical nephrectomy: first experience with the novel Da Vinci single-site platform. *World J Urol*. 2014;32:273–276.
21. Buffi NM, Lughezzani G, Fossati N, et al. Robot-assisted, single-site, dismembered pyeloplasty for ureteropelvic junction obstruction with the new da Vinci platform: a stage 2a study. *Eur Urol*. 2015;67(1):151–156.

22. Kaouk JH, Haber GP, Autorino R, et al. A novel robotic system for single-port urologic surgery: first clinical investigation. *Eur Urol*. 2014;66(6):1033−1043.
23. Ramirez D, Maurice MJ, Kaouk JH. Robotic perineal radical prostatectomy and pelvic lymph node dissection using a purpose-built single-port robotic platform. *BJU Int*. 2016;118(5):829−833.
24. Maurice MJ, Kaouk JH. Robotic radical perineal cystectomy and extended pelvic lymphadenectomy: initial investigation using a purpose-built single-port robotic system. *BJU Int*. 2017;120(6):881−884.
25. Kaouk JH, Sagalovich D, Garisto J. Robot-assisted transvesical partial prostatectomy using a purpose-built single-port robotic system. *BJU Int*. 2018;122(3):520−524.
26. Garisto JD, Bertolo R, Kaouk J. Technique for docking and port placement using a purpose-built robotic system (SP1098) in human cadaver. *Urology*. 2018;119:91−96.
27. Kaouk J, Garisto J, Bertolo R. Robotic urologic surgical interventions performed with the single port dedicated platform: first clinical investigation. *Eur Urol*. 2019;75(4):684−691.
28. Lenfant L, Corrigan D, Beksac AT, Schwen Z, Kaouk J. Learning curve analysis of single-port robot-assisted extraperitoneal prostatectomy using the CUSUM method. *BJU Int*. 2021. Available from: https://doi.org/10.1111/bju.15588. Sep 6 Epub ahead of print. PMID: 34487399.
29. Rassweiler JJ, Autorino R, Klein J, et al. Future of robotic surgery in urology. *BJU Int*. 2017;120(6):822−841. Available from: https://doi.org/10.1111/bju.13851. Epub 2017 Apr 22. PMID: 28319324.
30. Namdarian B, Dasgupta P. What robot for tomorrow and what improvement can we expect? *Curr Opin Urol*. 2018;28(2):143−152.
31. Cisu T, Crocerossa F, Carbonara U, Porpiglia F, Autorino R. New robotic surgical systems in urology: an update. *Curr Opin Urol*. 2021;31(1):37−42.
32. Seeliger B, Diana M, Ruurda JP, Konstantinidis KM, Marescaux J, Swanström LL. Enabling single-site laparoscopy: the SPORT platform. *Surg Endosc*. 2019;33(11):3696−3703. Available from: https://doi.org/10.1007/s00464-018-06658-x. Epub 2019 Jan 8. PMID: 30623255; PMCID: PMC6795913.

CHAPTER

Applications of single-port robotic platforms in urology: an overview

2

Marcio Covas Moschovas[1,2] and Vipul Patel[1,2]
[1]*AdventHealth Global Robotics Institute, Urology, Celebration, FL, United States*
[2]*University of Central Florida (UCF), Orlando, FL, United States*

2.1 Introduction

The da Vinci SP robot was cleared by the Food and Drug Administration (FDA) in November 2018. Due to its short life on the market, studies describing the da Vinci SP (SP) approach to urologic surgery are very recent in the literature. The first study reporting the SP clinical application was described by Kaouk and colleagues, who performed 19 procedures, 11 radical prostatectomies, and 8 nephrectomies.[1] The author reported feasible procedures with no conversions to alternative surgical approaches. Since then, the use of the SP has expanded, and multiple authors have described other applications for several types of urological procedures.[2,3]

In this scenario, radical prostatectomy (RP) has become the most common surgical procedure performed with the SP robot in most centers, followed by kidney surgery.[3] However, due to limited follow-up time, the long-term outcomes are still unknown for SP patients. In this chapter, we will describe current urologic applications and the available outcomes of the SP platform.

2.2 Da Vinci SP settings

2.2.1 Instruments and pedals

The main modification of this platform regards the instrument's design and articulated scope working simultaneously in only one trocar. Instead of four independent trocars placed in different positions, the SP console has a single port housing three, biarticulated instruments, and flexible scope (Fig. 2.1). This platform was initially built with an extra pedal, named the relocation pedal, used to drive all instruments and scope to different quadrants at the same time (Fig. 2.2). Until now, all studies published in the literature were performed with this pedal configuration. However, the SP had a recent update, which removed the extra pedal and changed the relocation function to a hand command.[4]

Single-Port Robotic Surgery in Urology. DOI: https://doi.org/10.1016/B978-0-323-91906-7.00008-9

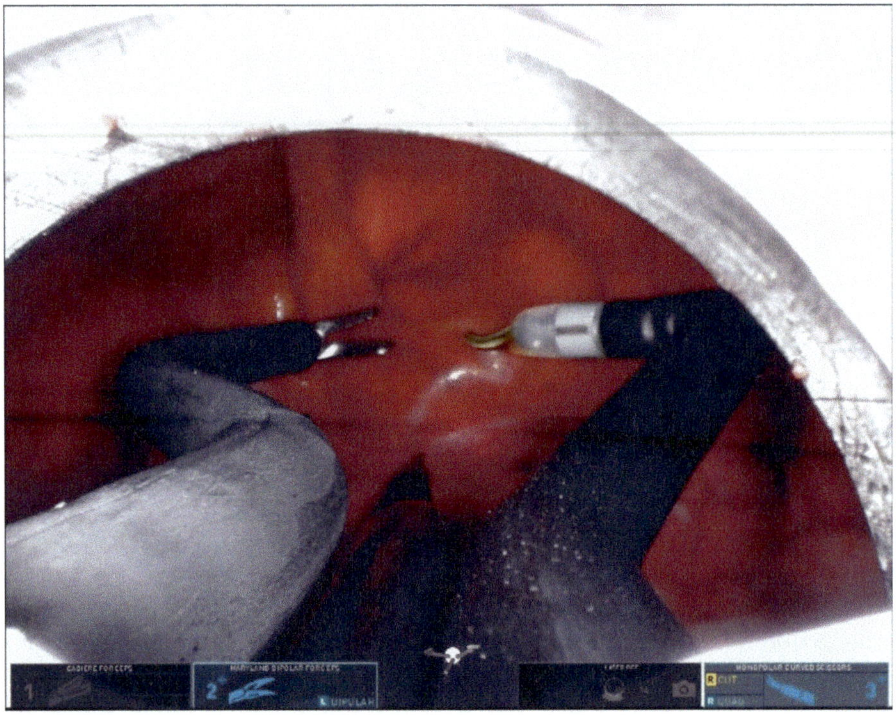

FIGURE 2.1

Biarticulated instruments seen inside the SP trocar.

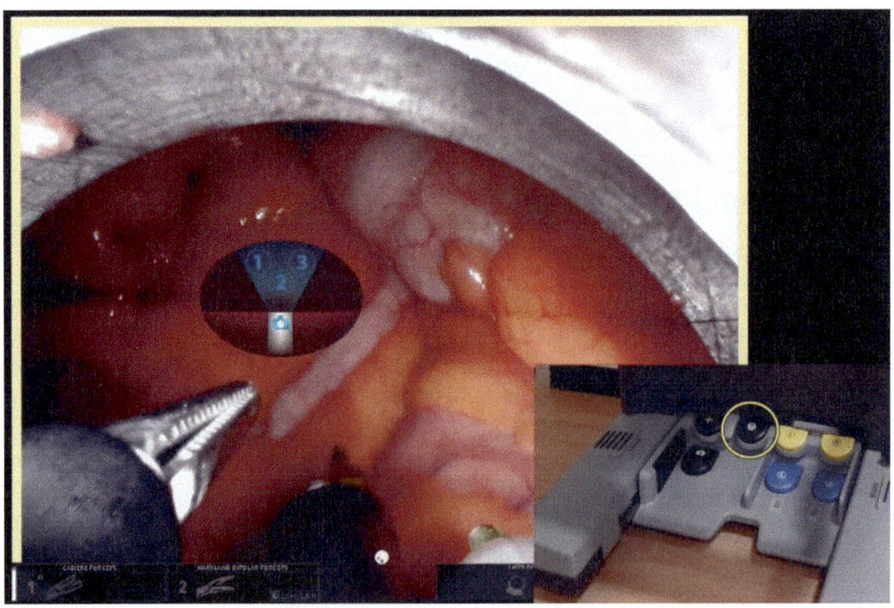

FIGURE 2.2

Relocation pedal.

2.2.2 Port placement

Different groups in the literature have described multiple techniques of SP trocar placement for RP and other urologic procedures. Appropriate port placement is necessary to allow the biarticulated instruments to achieve optimal working distance and triangulation. Placing the instruments too close to the surgical site increases internal clashing and restricts the scope visualization. This can be overcome by using the floating trocar during extraperitoneal SP-RARP and kidney surgery, because even with an infraumbilical incision, the trocar position (outside of the body) will provide the appropriate working distance.

2.2.3 Challenges and limitations

After achieving the learning curve with the SP platform, we identified factors that may prove to be challenges during the surgical procedure. The scope and instrument triangulation demand a learning curve in terms of angles and working distances; all arms are connected on the same base and have to work in the same space. The instruments are constantly repositioned during the surgical steps and the flexible scope angles are modified accordingly to provide optimal view and avoid clashing with the other instruments (Fig. 2.3).

The SP instruments have a reduced caliber (6 mm) compared to the XI (8 mm). Therefore this modification affects the capacity of traction, gripping, and dissection. By having a more delicate instrument, this robot is appropriate for selected cases due to its delicate tips and lack of rigid arms to provide traction when needed.

2.3 Preclinical studies

The first studies with the SP console were performed in cadavers using a prototype platform called SP999.[3] Table 2.1 summarizes these preclinical findings. The authors described different types of procedures with no conversions to open surgery or other robotic approach.[5–13]

2.4 SP applications in urologic surgeries

2.4.1 Kidney and ureters

The SP approach to kidney surgery has also been expanding in different centers, especially using extraperitoneal access. Kaouk and colleagues reported the surgical outcomes for three patients with renal masses who underwent SP partial nephrectomy (PN). No conversions or intraoperative complications were reported in this series. The group described a median operative time of 180 minutes,

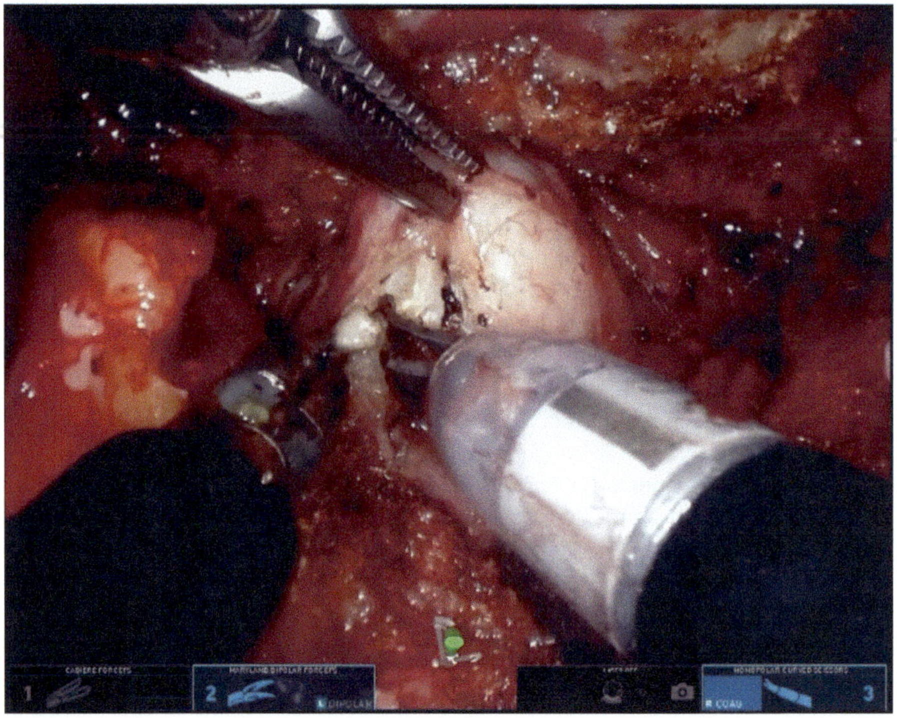

FIGURE 2.3

Cadiere, bipolar, and scissors positioned during the seminal vesicles dissection.

average ischemia time of 25 minutes, and 180 mL of blood loss (average). One patient had postoperative bleeding and underwent arterial embolization.[14] In a different series, Frang et al. described the SP technique for PN and radical nephrectomy in a cohort of 16 patients with cortical tumors.[15]

The SP platform has also been described for treating benign kidney conditions. Heo and colleagues described the outcomes of SP pyeloplasty in three patients with ureteropelvic junction obstruction. The author used the Anderson-Haynes technique in all patients and the transperitoneal approach with no complications or conversions. The entire procedure was performed with only one incision using the GelPOINT to allocate the robotic trocar and eventual suction and assistant.[16] In this scenario, Agarwal et al. described the SP approach to adult pyeloplasty.[17]

SP ureteric reimplantation for benign ureteric stricture was described by Kaouk and colleagues in three patients. The first two procedures used an assistant trocar (SP plus one) and a supraumbilical GelPOINT to house the SP trocar. No intraoperative complications were reported.[18] Hebert et al. also reported a ureteroneocystostomy in one patient with right distal ureteral stricture.[19]

Table 2.1 Preclinical studies.

Author	Year	Type of surgery	Number of cases	Type of access	Intraoperative complications	Median operative time (min)
Ramirez et al.[5]	2016	Perineal radical prostatectomy	1	Transperitoneal	None	180–240
		Lymphadenectomy	2			
Maurice et al.[6]	2017	Radical nephrectomy	2	Retroperitoneal	None	91.8
		Partial nephrectomy	4			100
Maurice et al.[7]	2017	Radical cystectomy + LND	2	Transperitoneal	None	197 and 200
Kaouk et al.[8]	2018	Partial prostatectomy	3	Transvesical	None	43
Garisto et al.[10]	2018	Transperitoneal intracorporeal Ileal conduit	1	Transperitoneal	None	90
Garisto et al.[12]	2018	Port placement for different types of surgeries	14	Transperitoneal Retroperitoneal Transvesical	None	N/A
Ng et al.[9]	2019	Radical prostatectomy	1	Retzius	None	146
Garisto et al.[11]	2019	Radical cystectomy	1	Perineal	None	185
Garisto et al.[13]	2019	Renal transplant	1	Transperitoneal	None	182

Furthermore, three different procedures were described by Billah and colleagues in six patients. Four patients underwent pyeloplasty, one patient underwent ureteroplasty, and one patient underwent ureterectomy. The author has described feasible procedures with no intraoperative complications.[20]

Finally, the initial experience of SP kidney transplantation was described by Kaouk and colleagues as an alternative for robotic transplantation. The author has described nine cases with no complications and satisfactory renal function during the follow-up.[21]

2.4.2 Prostate

RP is the most common procedure performed with the SP robot (Fig. 2.4). The outcomes in the current literature are presented in Table 2.2.

Five different accesses have been described for approaching RP with the SP robot. The first SP–RP reports were performed with transperitoneal access. Then, some modifications were described, and other techniques were reported using extraperitoneal, Retzius-sparing, transperineal, and transvesical approaches. In

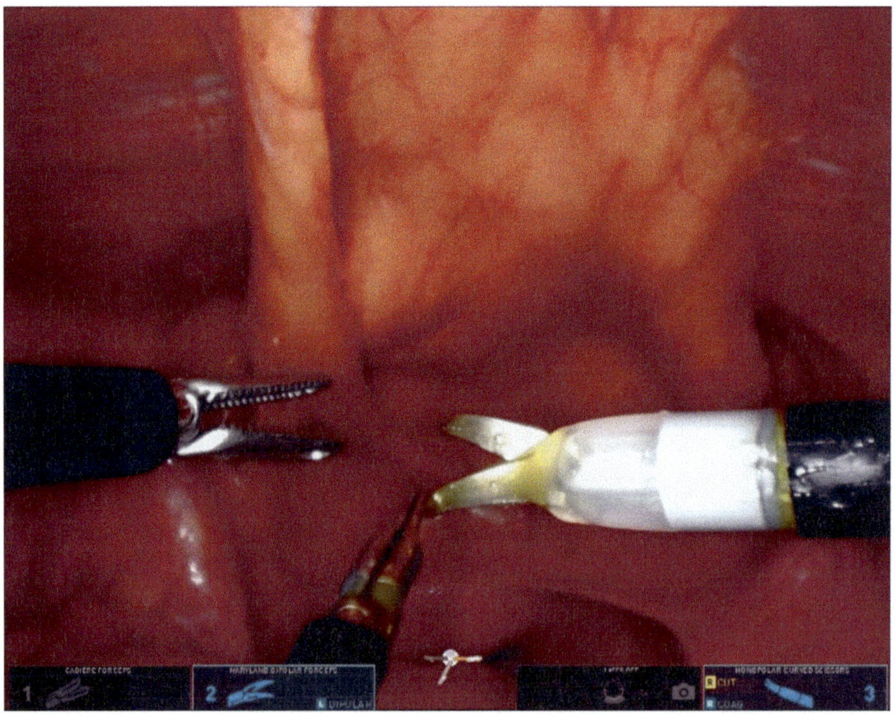

FIGURE 2.4

Transperitoneal SP robotic-assisted radical prostatectomy.

Table 2.2 Current literature reporting radical prostatectomy.

Author	Year	Number of cases	Access	Intraoperative complication rates	Postoperative complication rates (Clavien)	Operative time (min)	Console time (min)	Blood loss (mL)	Positive margins (%)
Kaouk et al.[1]	2014	11	TPR	None		239	NA	350	18
Kaouk et al.[22]	2019	3	TPR	None	None	180–300	NA	50–100	33
Ng et al.[23]	2019	20	TPR	None	25% (Clavien 1 and 2)	208	NA	296	55
Agarwal et al.[24]	2019	49	TPR or Retzius	None	8.1% (Clavien1)	161	NA	200	28
Kaouk et al.[25]	2019	10	RTP	None	None	197	NA	50–400	50
Dobbs et al.[26]	2019	10	TPR	None	None	234	189	20–150	20
Kaouk et al.[22]	2019	5	TPR	None	25% (Clavien 2)	180–330	NA	50–750	20
Dobbs et al.[26]	2019	24	TPR	1 Bowel serosal injury	45% (Clavien 1 to 4B)	191–343	NA	75	
Kaouk et al.[27]	2019	46 52	TPR EPT	None None	15% (Clavien1 and 2) 11% (Clavien1 and 2)	201	NA	117	27
Covas Moschovas et al.[28]	2020	26	TPR	None	None	121	85	50	11
Kaouk et al.[29]	2020	8 52	TPR EPT	None	None	106–281	NA	50–200	30
Valero et al.[30]	2020	1	TPR	None	None	256	108	100	0

(*Continued*)

Table 2.2 Current literature reporting radical prostatectomy. *Continued*

Author	Year	Number of cases	Access	Intraoperative complication rates	Postoperative complication rates (Clavien)	Operative time (min)	Console time (min)	Blood loss (mL)	Positive margins (%)
Kim et al.[31]	2020	20	TPR	None	None	200–255	165–210	155–300	35
Jones et al.[32]	2020	23	TPR	None	26%	236	NA	50	39
Wilson et al.[33]	2020	60	EPT	None	18% (Clavien 3a)	198	NA	179	23
Vigneswaran et al.[34]	2020	45 5	TPR EPT	None	14%	290	230	100	40
Covas Moschovas et al.[35]	2021	50	TPR	None	None	118	80	50	14
Abaza et al.[36]	2021	59	TPR	NA	NA	NA	NA	NA	NA
Covas Moschovas et al.[37]	2021	71	TPR	None	None	114	80	55	17
Talamini et al.[38]	2021	20	TPR	1 Bowel serosal injury	20% (Clavien 1–4B)	225	191	20–250	45
Lenfant et al.[39]	2021	26	RPP	None	50% (Clavien1–3A)	255	NA	100	65.4
Kaouk et al.[40]	2021	20	TVS	None	5% (Clavien 1)	199	119	135	15

EPT, *extraperitoneal;* NA, *nonavailable data;* RPP, *radical perineal prostatectomy;* TNP, *transperitoneal;* TVS, *transvesical.*

summary, the intraoperative complication rates range from 0% to 5%, with bowel serosal injury being the most common complication. The postoperative complications range from 0% to 45%, with reports ranging from Clavien 1 to Clavien 4B. Total operative time ranges from 114 to 343 minutes, and console time from 80 to 210 minutes. Blood loss ranges from 50 to 750 mL. Positive surgical margins range from 0% to 65.4%.

Only a few authors have compared the SP and multiport approaches to RP. Some described early discharge and less narcotic use in favor of the SP surgery, while others described similar postoperative outcomes in terms of hospital stay and pain scores.[34,35,39,41] However, all series published to date are based on retrospective studies and have their inherent risk of bias.

Simple prostatectomy with the SP was described by Steinberg et al. in a series of 10 cases using the extraperitoneal access for patients with prostates larger than 80 g. The author reported successful procedures with no conversions or intraoperative complications.[42] In this scenario, Kaouk and colleagues also reported their SP transvesical simple prostatectomy experience in 10 patients with benign prostatic hyperplasia. No intraoperative complications were reported, and only one patient had epididymitis in the postoperative period.[43]

2.4.3 Bladder

The clinical application of the SP robot in bladder surgery was reported by Kaouk et al. after performing radical cystectomy with lymphadenectomy in four patients using a single port plus one trocar placement. Three patients (75%) underwent intracorporeal urinary diversion, and only one patient had a postoperative complication (Clavien 1).[44] Zhang and colleagues also reported their experience with SP radical cystectomy in four patients with no intraoperative complications.[45]

Gross and colleagues compared the SP and multiport robotic approaches to radical cystectomy, describing a higher lymph node yield for the multiport group and no statistical difference in terms of operative time, blood loss, complication rates, readmissions, and positive surgical margins.[46]

2.4.4 Reconstructive surgery

Some groups have described the SP application in reconstructive procedures. Acar and colleagues reported nine vaginoplasties using the Davydov technique with no intraoperative complications. Two patients had 3B complications and were readmitted in the follow-up period.[47]

Furthermore, three patients who underwent SP sacrocolpopexy were described by Ganesan and colleagues. During the procedures, the authors used magnetic tissue retraction. All procedures underwent without complications, with an operative time ranging from 198 to 247 minutes and maximum blood loss of 50 mL.[48]

2.4.5 Pediatric

Kang and colleagues reported the first pediatric procedure performed with the da Vinci SP describing a pyeloplasty using the Anderson-Hynes technique. As a result, a 10-year-old patient with antenatal hydronephrosis had surgical treatment for ureteropelvic junction obstruction with no intraoperative complications.[49]

Parikh et al. described the step-by step technique for SP Mitrofanoff in a 14-year-old male with spinal injury. The author described a successful procedure with a median operative time of 120 minutes and minimum blood loss.[50]

2.4.6 Surgical treatment for infected lymphoceles

Covas Moschovas and colleagues also reported the SP approach to infected lymphocele drainage in patients with a previous refractory radiologic intervention. The author described three patients who underwent a successful drainage with no recurrence during the study follow-up.[51]

2.4.7 Studies reporting several types of procedures

The pioneering study reporting the SP use in the literature was performed by Kaouk and colleagues. The author had described the initial outcomes for 19 surgeries. Eleven patients underwent RP, four underwent PN, two underwent radical nephrectomy, and two underwent simple nephrectomy. No intraoperative complications were reported. However, postoperative complications ranged from 0% to 45% among the different procedures.[1] Afterward, the same group had described other studies describing the intraoperative and short-term outcomes for the SP surgery performing RP, PN, radical nephrectomy, pyeloplasty, radical cystectomy, and ureteral reimplantation.[22]

Doobs et al. also described the short-term outcomes of 45 patients who underwent SP surgery for prostate, kidney, and bladder diseases.[26] Finally, Valero and colleagues reported the technique combining left PN and RP in the same patient. No intraoperative complications were reported.[30]

2.5 Conclusion

The da Vinci SP role in urologic procedures has been expanding and improving in multiple centers. SP surgery is feasible and safe for the surgical treatment of malignant and benign genitourinary diseases in the urology field. RP is still the most common surgery performed with this robot. However, special attention must be taken when comparing the outcomes of this platform with the previous multiport series because the current SP literature is still based on retrospective studies and its inherent risk of bias.

References

1. Kaouk JH, Haber GP, Autorino R, et al. A novel robotic system for single-port urologic surgery: first clinical investigation. *Eur Urol*. 2014;66(6):1033−1043. Available from: https://doi.org/10.1016/j.eururo.2014.06.039. Epub 2014 Jul 17. PMID: 25041850.
2. Covas Moschovas M, Seetharam Bhat KR, Onol FF, et al. Single-port technique evolution and current practice in urologic procedures. *Asian J Urol*. 2021;8(1):100−104. Available from: https://doi.org/10.1016/j.ajur.2020.05.003. Epub 2020 May 22. PMID: 33569276; PMCID: PMC7859361.
3. Covas Moschovas M, Bhat S, Rogers T, et al. Applications of the da Vinci single port (SP) robotic platform in urology: a systematic literature review. *Minerva Urol Nephrol*. 2021;73(1):6−16. Available from: https://doi.org/10.23736/S0393-2249.20.03899-0. Epub 2020 Sep 29. PMID: 32993277.
4. Covas Moschovas M, Bhat S, Rogers T, et al. Da Vinci SP platform updates and modifications: the first impression of new settings. *J Robot Surg*. 2021;15(6):977−979. Available from: https://doi.org/10.1007/s11701-021-01248-x. Epub ahead of print. PMID: 34053019.
5. Ramirez D, Maurice MJ, Kaouk JH. Robotic perineal radical prostatectomy and pelvic lymph node dissection using a purpose-built single-port robotic platform. *BJU Int*. 2016;118:829−833.
6. Maurice MJ, Ramirez D, Kaouk JH. Robotic laparoendoscopic single-site retroperitoneal renal surgery: initial investigation of a purpose-built single-port surgical system. *Eur Urol*. 2017;71:643−647.
7. Maurice MJ, Kaouk JH. Robotic radical perineal cystectomy and extended pelvic lymphadenectomy: initial investigation using a purpose-built single-port robotic system. *BJU Int*. 2017;120:881−884.
8. Kaouk J, Sagalovich D, Garisto J. Robotic transvesical partial pros-tatectomy using a purpose-built single port robotic system. *BJU Int*. 2018;122:520−524. Available from: https://doi.org/10.1111/bju.14194.
9. Ng C-F, Chan ESY, Teoh JYC. The use of the da Vinci SP system for retzius-sparing radical prostatectomy in cadaveric model. *Urology*. 2019;125:260. Available from: https://doi.org/10.1016/j.urology.2018.11.042.
10. Garisto J, Bertolo R, Chan E, Kaouk J. Single-port trans-perineal approach to cysto-prostatectomy with intracorporeal ileal conduit urinary diversion and lymph-nodes dissection using a purpose-built robotic system: surgical steps in a preclinical model. *Int Braz J Urol*. 2019;45(4):854−855. Available from: https://doi.org/10.1590/S1677-5538.IBJU.2018.0524.
11. Garisto J, Bertolo R, Kaouk J. Trans-perineal approach for intracorporeal ileal conduit urinary diversion using a purpose-built single-port robotic system: step-by-step. *Urology*. 2018;122:179−184. Available from: https://doi.org/10.1016/j.urology.2018.08.019.
12. Garisto J, Bertolo R, Kaouk J. Technique for docking and port placement using a purpose-built robotic system (SP1098) in human cadaver. *Urology*. 2018;119:91−96. Available from: https://doi.org/10.1016/j.urology.2018.05.036.
13. Garisto J, Eltemamy M, Bertolo R, Miller E, Wee A, Kaouk J. Single port robot-assisted transperitoneal kidney transplant using the SP® surgical system in a preclinical model. *Int Braz J Urol*. 2020;46(4):680−681. Available from: https://doi.org/10.1590/S1677-5538.IBJU.2019.191.

14. Kaouk J, Garisto J, Eltemamy M, Bertolo R. Pure single-site robot-assisted partial nephrectomy using the SP surgical system: initial clinical experience. *Urology*. 2019;124:282–285. Available from: https://doi.org/10.1016/j.urology.2018.11.024.
15. Fang AM, Saidian A, Magi-Galluzzi C, Nix JW, Rais-Bahrami S. Single-port robotic partial and radical nephrectomies for renal cortical tumors: initial clinical experience. *J Robot Surg*. 2020;14:773–780. Available from: https://doi.org/10.1007/s11701-020-01053-y.
16. Heo JE, Kang SK, Koh DH, et al. Pure single-site robot-assisted pyeloplasty with the da Vinci SP surgical system: initial experience. *Investig Clin Urol*. 2019;60(4):326–330. Available from: https://doi.org/10.4111/icu.2019.60.4.326.
17. Agarwal DK, Hebert KJ, Gettman MT, Viers BR. How to perform a robotic pyeloplasty utilizing the da Vinci SP platform: tips and tricks. *Transl Androl Urol*. 2020;9 (2):919–924. Available from: https://doi.org/10.21037/tau.2019.11.08.
18. Kaouk JH, Garisto J, Eltemamy M, Bertolo R. Robot-assisted surgery for benign distal ureteral strictures: step-by-step technique using the SP® surgical system. *BJU Int*. 2019;123:733–739. Available from: https://doi.org/10.1111/bju.14635.
19. Hebert K.J., Joseph J., Gettman M., Tollefson M., Frank I., Viers B.R. Technical considerations of single port ureteroneocystostomy utilizing da Vinci SP platform. *Urology* 129:236. https://doi.org/10.1016/j.urology.2019.03.020.
20. Billah MS, Stifelman M, Munver R, Tsui J, Lovallo G, Ahmed M. Single port robotic assisted reconstructive urologic surgery—with the da Vinci SP surgical system. *Transl Androl Urol*. 2020;9(2):870–878. Available from: https://doi.org/10.21037/tau.2020.01.06.
21. Kaouk J, Eltemamy M, Aminsharifi A, et al. Initial experience with single-port robotic-assisted kidney transplantation and autotransplantation. *Eur Urol*. 2021;80 (3):366–373. Available from: https://doi.org/10.1016/j.eururo.2021.03.002.
22. Kaouk J, Garisto J, Bertolo R. Robotic urologic surgical interventions performed with the single port dedicated platform: first clinical investigation. *Eur Urol*. 2019;75 (4):684–691. Available from: https://doi.org/10.1016/j.eururo.2018.11.044. Epub 2018 Dec 3. PMID: 30522914.
23. Ng CF, Teoh JY, Chiu PK, et al. Robot-assisted single-port radical prostatectomy: a phase 1 clinical study. *Int J Urol*. 2019;26(9):878–883. Available from: https://doi.org/10.1111/iju.14044. Epub 2019 Jun 30. PMID: 31257704.
24. Agarwal DK, Sharma V, Toussi A, et al. Initial experience with da Vinci Single-port robot-assisted radical prostatectomies. *Eur Urol*. 2020;77(3):373–379. Available from: https://doi.org/10.1016/j.eururo.2019.04.001. Epub 2019 Apr 19. PMID: 31010600.
25. Kaouk J, Valero R, Sawczyn G, Garisto J. Extraperitoneal single-port robot-assisted radical prostatectomy: initial experience and description of technique. *BJU Int*. 2020;125(1):182–189. Available from: https://doi.org/10.1111/bju.14885.
26. Dobbs RW, Halgrimson WR, Talamini S, Vigneswaran HT, Wilson JO, Crivellaro S. Single-port robotic surgery: the next generation of minimally invasive urology. *World J Urol*. 2020;38(4):897–905. Available from: https://doi.org/10.1007/s00345-019-02898-1. Epub 2019 Aug 28. PMID: 31463560.
27. Jihad K, Alireza A, Clark AW, et al. Extraperitoneal vs transperitoneal single-port robotic radical prostatectomy: a comparative analysis of perioperative outcomes. *J Urol*. 2020;203(6):1135–1140. Available from: https://doi.org/10.1097/JU.0000000000000700.
28. Covas Moschovas M, Bhat S, Rogers T, et al. Technical modifications necessary to implement the da Vinci Single-port robotic system. *Eur Urol*. 2020;78(3):415–423. Available from: https://doi.org/10.1016/j.eururo.2020.01.005.

29. Kaouk J, Aminsharifi A, Sawczyn G, et al. Single-port robotic urological surgery using purpose-built single-port surgical system: single-institutional experience with the first 100 cases. *Urology*. 2020;140:77–84. Available from: https://doi.org/10.1016/j.urology.2019.11.086.
30. Valero R, Sawczyn G, Garisto J, Yau R, Kaouk J. Multiquadrant combined robotic radical prostatectomy and left partial nephrectomy: a combined procedure by a single approach. *Actas Urol Esp*. 2020;44(2):119–124. Available from: https://doi.org/10.1016/j.acuro.2019.06.004.
31. Kim KH, Song W, Yoon H, Lee DH. Single-port robot-assisted radical prostatectomy with the da Vinci SP system: a single surgeon's experience. *Investig Clin Urol*. 2020;61(2):173–179. Available from: https://doi.org/10.4111/icu.2020.61.2.173.
32. Jones R, Dobbs RW, Halgrimson WR, et al. Single port robotic radical prostatectomy with the da Vinci SP platform: a step by step approach. *Can J Urol*. 2020;27(3):10263–10269.
33. Wilson CA, Aminsharifi A, Sawczyn G, et al. Outpatient extraperitoneal single-port robotic radical prostatectomy. *Urology*. 2020;144:142–146. Available from: https://doi.org/10.1016/j.urology.2020.06.029. Epub 2020 Jun 30. PMID: 32619598.
34. Vigneswaran HT, Schwarzman LS, Francavilla S, Abern MR, Crivellaro S. A comparison of perioperative outcomes between single-port and multiport robot-assisted laparoscopic prostatectomy. *Eur Urol*. 2020;77(6):671–674. Available from: https://doi.org/10.1016/j.eururo.2020.03.031. Epub 2020 Apr 9. PMID: 32279904.
35. Covas Moschovas M, Bhat S, Onol F, Rogers T, Patel V. Early outcomes of single-port robot-assisted radical prostatectomy: lessons learned from the learning-curve experience. *BJU Int*. 2021;127(1):114–121. Available from: https://doi.org/10.1111/bju.15158. Epub 2020 Aug 2. PMID: 32623822.
36. Abaza R, Murphy C, Bsatee A, Brown Jr DH, Martinez O. Single-port robotic surgery allows same-day discharge in majority of cases. *Urology*. 2021;148:159–165. Available from: https://doi.org/10.1016/j.urology.2020.08.092. Epub 2020 Nov 17. PMID: 33217457.
37. Covas Moschovas M, Bhat S, Sandri M, et al. Comparing the approach to radical prostatectomy using the multiport da Vinci Xi and da Vinci SP robots: a propensity score analysis of perioperative outcomes. *Eur Urol*. 2021;79(3):393–404. Available from: https://doi.org/10.1016/j.eururo.2020.11.042. Epub 2020 Dec 24. PMID: 33357994.
38. Talamini S, Halgrimson WR, Dobbs RW, Morana C, Crivellaro S. Single port robotic radical prostatectomy vs multi-port robotic radical prostatectomy: a human factor analysis during the initial learning curve. *Int J Med Robot*. 2021;17(2):e2209. Available from: https://doi.org/10.1002/rcs.2209. Epub 2020 Dec 29. PMID: 33320437.
39. Lenfant L, Garisto J, Sawczyn G, et al. Robot-assisted radical prostatectomy using single-port perineal approach: technique and single-surgeon matched-paired comparative outcomes. *Eur Urol*. 2021;79(3):384–392. Available from: https://doi.org/10.1016/j.eururo.2020.12.013. Epub 2020 Dec 21. PMID: 33357990.
40. Kaouk J, Beksac AT, Zeinab MA, Duncan A, Schwen ZR, Eltemamy M. Single port transvesical robotic radical prostatectomy: initial clinical experience and description of technique. *Urology*. 2021;155:130–137. Available from: https://doi.org/10.1016/j.urology.2021.05.022.
41. Noël J, Covas Moschovas M, Sandri M, et al. Patient surgical satisfaction after da Vinci® single-port and multi-port robotic-assisted radical prostatectomy: propensity score-matched analysis. *J Robot Surg*. 2021;1–9. Available from: https://doi.org/10.1007/s11701-021-01269-6. Jun 18.

42. Steinberg RL, Passoni N, Garbens A, et al. Initial experience with extraperitoneal robotic-assisted simple prostatectomy using the da Vinci SP surgical system. *J Robotic Surg*. 2020;14(4):601–607. Available from: https://doi.org/10.1007/s11701-019-01029-7.
43. Kaouk J, Sawczyn G, Wilson C, et al. Single-port percutaneous transvesical simple prostatectomy using the SP® robotic system: initial clinical experience. *Urology*. 2020;141:173–177. Available from: https://doi.org/10.1016/j.urology.
44. Kaouk J, Garisto J, Eltemamy M, Bertolo R. Step-by-step technique for single-port robot-assisted radical cystectomy and pelvic lymph nodes dissection using the da Vinci® SP™ surgical system. *BJU Int*. 2019. Available from: https://doi.org/10.1111/bju.14744.
45. Zhang M, Thomas D, Salama G, Ahmed M. Single port robotic radical cystectomy with intracorporeal urinary diversion: a case series and review. *Transl Androl Urol*. 2020;9(2):925–930. Available from: https://doi.org/10.21037/tau.2020.01.19.
46. Gross JT, Vetter JM, Sands KG, et al. Initial experience with single-port robot-assisted radical cystectomy: comparison of perioperative outcomes between single-port and conventional multiport approaches. *J Endourol*. 2021;35:1177–1183. Available from: https://doi.org/10.1089/end.2020.1227. Apr 5.
47. Acar O, Sofer L, Dobbs RW, et al. Single port and multiport approaches for robotic vaginoplasty with the Davydov technique. *Urology*. 2020;138:166–173. Available from: https://doi.org/10.1016/j.urology.2019.11.043.
48. Ganesan V, Goueli R, Rodriguez D, Hess D, Carmel M. Single-port robotic-assisted laparoscopic sacrocolpopexy with magnetic retraction: first experience using the SP da Vinci platform. *J Robotic Surg*. 2020;14. Available from: https://doi.org/10.1007/s11701-020-01050-1.
49. Kang SK, et al. Robot-assisted laparoscopic single-port pyeloplasty using the da Vinci SP® system: initial experience with a pediatric patient. *J Pediatr Urol*. 2019;15(5):576–577.
50. Parikh N, Findlay B, Boswell T, Granberg C, Gargollo P. Single-port robotic Mitrofanoff in a pediatric patient. *J Pediatr Urol*. 2021;17(3):424–425. Available from: https://doi.org/10.1016/j.jpurol.2021.03.021. Epub ahead of print. PMID: 33865707.
51. Covas Moschovas M, Helman T, Reddy S, et al. Minimally invasive lymphocele drainage using the da Vinci® Single Port platform: step-by-step technique of a prostate cancer referral center. *J Endourol*. 2021. Available from: https://doi.org/10.1089/end.2020.1175. Mar 16.

CHAPTER

Single-port robot-assisted radical prostatectomy: transperitoneal versus extraperitoneal approach

3

Grace Chen and Simone Crivellaro

Department of Urology, University of Illinois at Chicago, Chicago, IL, United States

3.1 Introduction

Since the advent of the da Vinci Surgical System (Intuitive Surgical, Sunnyvale, California, United States) in 2000, the robotic-assisted radical prostatectomy (RARP) has quickly risen to become the gold standard for the minimally invasive surgical treatment of prostate cancer. The introduction of the da Vinci single-port (SP) system, which was approved by the FDA in 2018, increases the variety of approaches to the SP RARP described in the literature, with two of the most common approaches being transperitoneal and extraperitoneal. In this chapter, we will describe the major similarities and differences between the transperitoneal and extraperitoneal approaches with the SP, patient selection considerations and contraindications, potential complications, and preliminary outcomes data for each approach.

3.2 Basic technique of the transperitoneal SP RARP

The transperitoneal approach is currently the most popular approach for the multi-arm RARP and was therefore the first to be performed with the novel SP platform. The first report of the transperitoneal SP RARP being performed on a patient was published by Kaouk et al. in 2019, with a case report describing their surgical experience with two patients. This was closely followed by case series published by other academic centers, including Dobbs et al. in 2019.

The technique of the transperitoneal RARP can be summarized as follows. The patient is first placed in supine position with the arms tucked. Once the patient is prepped and draped in the usual sterile fashion and placed in steep Trendelenburg, a 3 cm vertical incision is made 3 cm superior to the umbilicus and the peritoneum is entered under direct vision via a Hasson technique. Once the peritoneal cavity is entered, the Alexis retractor (Applied Medical, Rancho Santa Margarita, California, United States) is introduced through the incision and

Single-Port Robotic Surgery in Urology. DOI: https://doi.org/10.1016/B978-0-323-91906-7.00006-5

secured under the peritoneum to provide 360 degree retraction of the incision. The GelPOINT Mini Advanced Access Platform (Applied Medical, Rancho Santa Margarita, California, United States) is then attached to the retractor. A 5 mm AirSeal (ConMed Corp., Utica, New York, United States) "side car" port is then placed through the same skin incision, through a different fascial incision and eventually through the Alexis retractor under direct digital control (see Fig. 3.1).

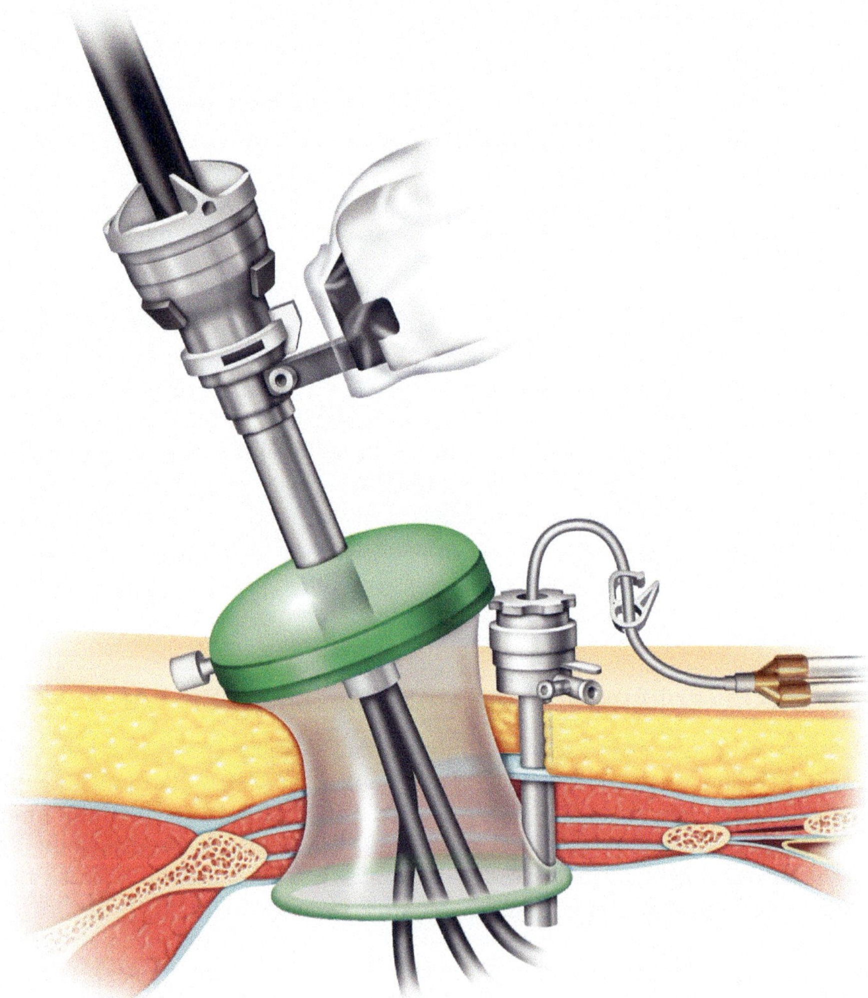

FIGURE 3.1

GelPOINT Mini system with 5 mm "side car" port inserted through plastic of Alexis retractor.

The robot is then docked to the SP access port with the camera at the 12 o'clock position, a monopolar scissor at 3 o'clock, a Cadiere forceps at 6 o'clock, and a fenestrated bipolar forceps at 9 o'clock.

The dissection is then undertaken with similar technique to a multiport transperitoneal RARP, with either a posterior or anterior approach to the dissection of the seminal vesicles depending on needs of the case.

Of note, our institution utilizes the remotely operated suction irrigation system (ROSI) (Vascular Technology, Nashua, New Hampshire, United States) for suction, which allows the surgeon at the console control of the suction throughout the procedure. This is placed through the "side car" port and switched out for instruments depending on operative need. Given the use of the Airseal system, the suction is able to be kept on for most of the procedure, thanks to the possibility of locking the suction through the ROSI pedal. This will allow instantaneous evacuation of smoke during the procedure, removing the need for the operator to hit the pedal for suction.

3.3 Basic technique of the extraperitoneal SP RARP

The first described extraperitoneal SP RARP was published by Kaouk et al. in 2019, with a handful of studies published since detailing single-institutional experiences with this approach.[1,2] The extraperitoneal technique for the multiport system is well described, and lends itself particularly well to the SP platform due to the small amount of working space required by the SP to be effective in the space of Retzius. The small working distance between the instruments and the camera and the overall smaller field of view makes the SP platform ideal for this approach. In addition, due to the single entry site needed for the SP approach, there is a reduced risk of accidental peritoneal puncture from lateral port placement, which is necessary for the multiport.[3,4]

Key differences to note with the SP extraperitoneal approach is that the patient can be positioned in the supine position without need for the steep Trendelenburg traditionally used for the transperitoneal approach. A single, horizontal 3 cm infraumbilical incision is made below the umbilicus, and electrocautery is used to divide the anterior rectus sheath in the midline. We then use blunt finger dissection to develop the space of Retzius to the pubic bone. Once the space of Retzius is developed just underneath the rectus fascia, we insert the Balloon Dissector Spacemaker (Covidien, Dublin, Ireland), which is inflated with 800 cc air to further expand the working space. A novel SP Access Port (Intuitive Surgical, Sunnyvale, California, United States) is used for the extraperitoneal approach. This access port consists of a combination of a wound retractor, an inflatable plastic globe that serves as a floating platform to provide extra space for the robotic arms, and an SP robotic trocar. The SP access port retractor is then introduced through the incision and secured under the posterior rectus fascia to provide 360 degree retraction of the incision. The SP access port itself is then attached to the retractor. A 5 mm "side car" port is then placed through the same skin incision, through a different fascial incision and eventually through the SP Access Port retractor under direct digital control. A 5 mm AirSeal (ConMed Corp., Utica, New York, United States) trocar is inserted for insufflation through the lateral blue entry point on the SP access port (see Fig. 3.2).

The extraperitoneal space is then insufflated and the robot is docked, with the robotic arms advanced into the space of Retzius under direct visualization. It must also be noted that the extraperitoneal approach is associated with increased carbon dioxide absorption, which may result in rising end-tidal CO_2 levels and hypercarbia which may result in systemic acidosis.[5] This will require careful monitoring by the anesthesiologist and sometimes adjustment in insufflation pressure by the surgical team. At our institution, we insufflate at an initial pressure of 8 mm Hg and we never go higher than 12 mm Hg. The use of the Airseal system is instrumental in keeping the pressure in this range. ROSI system is used for suction and is introduced through the "side car" port and kept on continuous suction. The camera is docked at the 12 o'clock position, the monopolar scissors at the 3 o'clock position, the fenestrated bipolar forceps at the 9 o'clock position, and

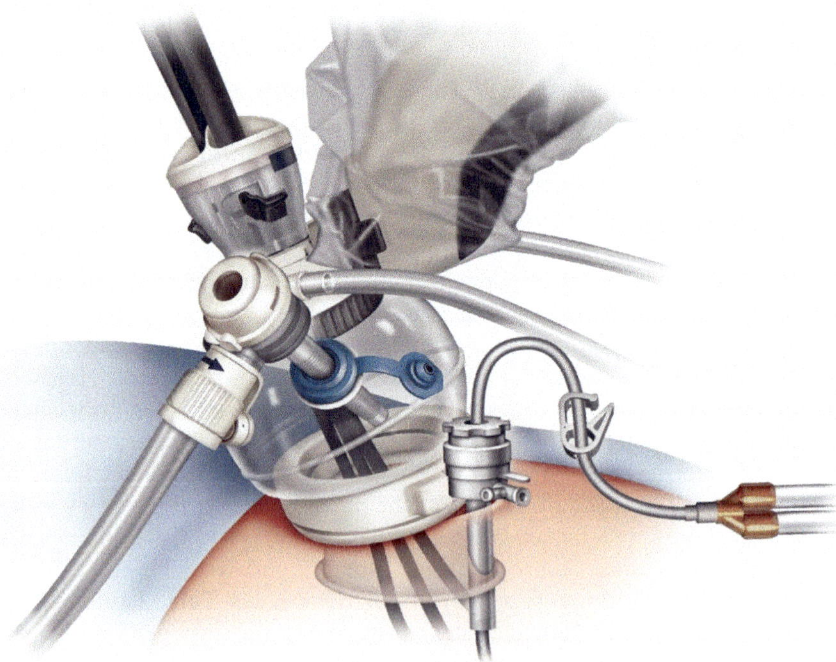

FIGURE 3.2

SP access port.

the Cadiere forceps at the 6 o'clock position. We then proceed with the procedure with the standard anterior approach to the multiport robotic radical prostatectomy. For the extraperitoneal SP RARP, the bladder does not need to be mobilized due to direct access to the prostate.

3.3.1 Patient selection: indications and contraindications

The indications for SP RARP are essentially the same as for open surgery or multiport surgery. The patients selected should have clinically localized prostate cancer that is either clinically confined to the prostate (stage T1 or T2) or extending beyond the margin of the prostate but still amenable to wide resection (T3).

Prior to undergoing SP RARP (similar to an open or multiport approach), one must understand the unique anatomical challenges posed by patients who have had prior complex pelvic surgeries such as partial colectomy or inguinal mesh herniorrhaphy, which may result in dense adhesions in the surgical field of interest. Although this is not an absolute contraindication to surgery, a transperitoneal approach can pose more challenges than an extraperitoneal access and therefore it can be considered as a contraindication for the former and an indication for the later.

One advantage of the SP system over the multiport is that making the extraperitoneal access more approachable there is less theoretical risk to morbidly obese patients. Traditionally, patients undergoing multiport transperitoneal RARP must be in a steep Trendelenburg position, which increases risk of nerve compression from positioning (particularly brachial plexus from strain on the shoulders) and potential respiratory compromise due to pressure on the diaphragm. The SP system allows for the procedure to be performed in more patients with the extraperitoneal approach, in the supine position with minimal Trendelenburg, minimizing the risk of these complications. Other well-documented risks of morbid obesity still hold, however, including risk of deep venous thrombosis and limited working space due to body habitus.

Absolute contraindications to either of the approaches include the inability to undergo general anesthesia due to severely reduced cardiopulmonary function or uncorrectable coagulopathy increasing risk of severe intraoperative bleeding.

It must be noted that a relative contraindication to the extraperitoneal approach is a history of prior laparoscopic extraperitoneal mesh herniorrhaphy, as this may cause dense adhesions in the space of Retzius, which significantly distorts normal anatomic landmarks and limits access to the retropubic space. For such patients, it may be preferable to proceed with a transperitoneal approach instead. This holds true for the SP platform as well as the multiport, although the SP extraperitoneal approach doesn't require lateral dissection, therefore allowing in most cases to access the prostate even when the extraperitoneal space has been already violated.

3.4 Complications

Some of the complications that can occur with extraperitoneal SP RARP are possible with the transperitoneal approach, including infection, bleeding potentially requiring blood transfusion, or other pelvic organ injury. Some other complications such as bowel injury or ileus are completely eliminated by the extraperitoneal approach, representing a big advantage.

The rate of postoperative fluid collections requiring percutaneous drainage such as lymphoceles, seromas, hematomas, or abscesses, and lymphoceles is comparable, but the management of this complication is significantly easier with the extraperitoneal approach, given the absence of contamination of the peritoneal cavity, hence the self containing course and the lesser risk of sepsis.

Overall though, there is not currently enough data to suggest that one approach is superior to another in terms of postoperative complications.[2]

3.5 Pros and cons of extraperitoneal versus transperitoneal approach

Given the novelty of the SP platform, only a handful of studies have been published thus far on SP RARP performed via either of the approaches. Multiple single center studies have been published, although only a single study has been published to date comparing the two approaches and outcomes. The data of demographics and outcomes from these single center studies is summarized in the following tables (Tables 3.1–3.4).

Overall, the argument in favor of the extraperitoneal approach is that it is a less invasive approach and allows RARP to be performed in hostile abdomens. As per Ref. 2, a significantly shorter postoperative hospital stay is associated with the extraperitoneal RARP (4.3 vs 25.7 hours, with same day discharge possible in the majority of extraperitoneal RARPs). This is surmised to be due to a number of factors, the first of which is the lack of entry into the peritoneum, which is known to be associated with peritoneal irritation and postoperative ileus resulting in a prolonged recovery. In addition, the lack of steep Trendelenberg and pneumoperitoneum may also result in lower rates of postoperative nausea and vomiting.[12] Due to direct access to the prostate, the extraperitoneal SP RARP may also reduce operative time compared to the transperitoneal approach (201 + 37.5 vs 248.2 + 42.3 minutes, $P < .00001$), resulting in less time under anesthesia, which is also associated with GI symptoms. For all of the above reasons, the extraperitoneal RARP is associated with fewer gastrointestinal symptoms in the immediate postoperative setting.[2]

There has also been noticed a significant difference in need for postoperative analgesia, with patients needing significantly less narcotics in the extraperitoneal group than transperitoneal.[2] There has been made a large impact on length of

Table 3.1 Transperitoneal—patient characteristics and demographics (Refs. [3,4,6–10]).

Author	Hospital	N	Avg Age (range)	BMI	Mean prostate size (mL)	Preop PSA	Gleason Score	Clinical T stage	Pathological T stage
Covas 2021	Adventhealth Global Robotics Institute	50	57	25	48	6.3	GS 6 - 15 (30%) GS 7 - 32 (66%) GS 8 to 10 - 2 (5%)	NR	NR
Aminsharifi 2020	Cleveland Clinic	120	64.3	29.3	NR	6.1	GS 6 - 15 (12.5%) GS 7 - 8 (66.7%) GS 8 to 10 - 25 (20.8%)	NR	T2 - 60 (50%) T3a,b - 60 (50%)
Abaza 2020	MD Anderson	34	61	27.1		7.8 (± 8.1)	GS 6 - 1 (3%) GS 7 - 31 (91%) GS 8 to 10 - 2 (2%)	T1c - 30 (88%) T2 - 4 (12%)	T0 - 0 Tx - 0 T2 - 22 (73%) T3a - 6 (20%) T3b - 2 (7%)
Abaza 2020	Ohiohealth Dublin Methodist Hospital	40	64	25.1		7.9 (± 6.6)	GS 6 - 7 (18%) GS 7 - 24 (60%) GS 8 to 10 - 9 (22%)	T1c - 24 (60%) T2 - 15 (38%) T3 - 1 (2%)	T0 - 1 (2.5%) Tx - 1 (2.5%) T2 - 18 (45%) T3a - 14 (35%) T3b - 6 (15%)
Ng 2019	Prince of Wales Hospital	20	67		36.6	15.3 (± 11.3)	GS 6 - 12 (60%) GS 7 - 4 (20%) GS 8 to 10 - 4 (20%)	T1c - 19 (95%) T3a - 1 (5%)	T2a - 3 (15%) T2b - 1 (5%) T2c - 7 (35%) T3a - 9 (45%)
Kim 2019	Ewha Womans University Seoul Hospital	20	66	24	32	NR	GS 6 - 1 (5%) GS 7 - 10 (50%) GS 8 to 10 - 9 - (45%)	NR	pT2 - 11 (55%) pT3a - 5 (25%) pT3b - 3 (15%) pT4 - 1 (5%)
Dobbs 2019	University of Illinois Hospital	10	52	29.4	55	11 (± 10.6)	GS 7.3 ± 1.2	NR	T1c - 1 (10%) T2a - 2 (20%) T2c - 3 (30%) T3a - 1 (10%) T3b - 3 (30%)
Agarwal 2019	Mayo Clinic	49	62	28.5	NR	6.4	GS 6 - 4 (9%) GS 7 - 37 (78%) GS 8 to 10 - 8 (13%)	cT1c - 35 (71%) cT2a/b/c - 8 (18%) cT3 - 5 (10%)	pT2 - 40 (85%) pT3a - 4 (9%) pT3b - 3 (6%)
Kaouk 2019	Cleveland Clinic	2	NR	NR	NR	NR	NR	NR	NR

Table 3.2 Transperitoneal—outcomes (Refs. [3,4,6–8,10]).

Author	Hospital	N	Average operative time (min)	LOS (days)	EBL (mL)	Pain score	Narcotic use	30 day readmit	Complication rate	Positive margins	Positive nodes (pN1)	PSA recurrence	Rate of SUI (>1 ppd)	Rate of ED
Covas 2021	Adventhealth Global Robotics Institute	50	118	0.6 days	50	2	4 (8%)	0 (0%)	0 (0%)	7 (14%)	0 (0%)	0 (0%)	11 (22%)	40 (80%)
Aminsharifi 2020	Cleveland Clinic	120	203	0.18 days	197	2	37/120 (30.8%)		16 (13.3%) Clavien 1-2 - 9 (7.5%) Clavien 3 - 7 (5.8%)	29 (24.2%)	[illegible]	8 [illegible]	[illegible]	NR
Abaza 2020	MD Anderson	34	179	1.1 days	124	NR	NR	1 (1.3%)	Clavien 1 - 1 (3%) Clavien 3a - 2 (6%)	17%	NR	NR	NR	NR
Abaza 2020	Ohiohealth Dublin Methodist Hospital	40	173	0.12 days	68	NR	NR	0	Clavien 1 - 1 (2.5%)	23%	NR	NR	NR	NR
Ng 2019	Prince of Wales Hospital	20	208	5 days	296	NR	NR	NR	Clavien 1 - 2 (10%) Clavien 2 - 3 (15%)		NR	5 (15%)	NR	NR
Kim 2019	Ewha Womans University Seoul Hospital	20	245	7 days	200	NR	NR	NR	Clavien >2 - 0 (0%)	7 (35%)	NR	1 (5%)	12 (60%)	16 (80%)
Dobbs 2019	University of Illinois Hospital	10	235	1 day (7 on POD#1, 3 on POD#2)	65	NR	NR	NR	0 (0%)	2 (20%)	NR	NR	NR	NR
Agarwal 2019	Mayo Clinic	49	161	1 day	200	NR	NR	NR	4 (8%) Clavien 2 - 4 (8%)	13 (28%)	2 (6%)	1 (3.6%)	5/21 (24%)	11/21 (52%)
Kaouk 2019	Cleveland Clinic	2	140	<1 day	"negligible"	NR	NR	NR	NR	NR	NR	NR	NR	NR

Table 3.3 Extraperitoneal—patient characteristics and demographics (Refs. [1,11]).

Author	Hospital	N	Avg Age (range)	BMI	Mean prostate size	Preop PSA	Gleason Score	Clinical T stage	Pathological T stage	Rate of ED
Wilson 2020	Cleveland Clinic	60	63 (47-75)	30 (19-41)	NR	7.6	GS 6 - 2 (3%) GS 7- 47 (78%) GS 8 to 10 - 9 (15%)	NR	pT2 - 32 (53.3%) >pT3a - 28 (46.7%)	NR
Kaouk 2020	Cleveland Clinic	10	62.3	30	NR	9	GS 6 - 1 (10%) GS 7 - 4 (40%) GS 8 to 10 - 5 (50%)	NR	NR	NR

Table 3.4 Extraperitoneal—outcomes (Refs. [1,11]).

Author	N	Average operative time (min)	LOS (days)	EBL (mL)	Avg pain score	Narcotic use	30 day readmit	Complication rate	Positive margins	Positive nodes (pN1)	PSA recurrence	Rate of SUI	Rate of ED
Wilson (2020)	60	198	0.18	179	2	52%	4 (6%)	Total - 11 (18%) Clavien 3a - 7 (11.6%)	14 (23%)	4 (6.7%)	NR	0-1 pads (28/58 , 48%) 2-3 pads (17/58, 29%) 4-5 pads (7/58, 12%) >5 pads (6/58, 10%)	SHIM 20-24 (1/19, 5%) SHIM 15-19 (1/19, 5%) SHIM 10-14 (2/19, 10%) SHIM 5-9 (13/19, 58%)
Kaouk (2020)	10	197	0	143		2 (20%)	0	0	5 (50%)	1 (10%)	NR	8 (80%) with SUI	8 (80%) with ED

hospital stay, with extraperitoneal SP RARP being able to be performed as an outpatient procedure with minimal to no opioid use and acceptable short-term functional and oncological outcomes.[1]

From an oncological standpoint the positive surgical margins have been comparable between the two groups, and from a functional standpoint there was no difference in either of the continence rate at 3 and 6 months.

One thing that must be noted is that historically the robotic extraperitoneal approach was generally associated with a lower lymph node yield than the transperitoneal approach. However, the single-port makes the pelvic lymphadenectomy more feasible, given the single access, the relocation feature and the flexible camera. Further study will clarify if the gap has been filled, thanks to this new approach.

3.6 Conclusion

Overall, the possibility of having in your armamentarium an extraperitoneal or transperitoneal approach for SP RARP gives the chance to make the procedure fit the patient and not the other way around. Additionally the extraperitoneal approach appears to have distinct advantages noted in the single retrospective study cited above, but of course further studies will be necessary to conclude that one approach is superior to the other.

References

1. Wilson, et al. Outpatient extraperitoneal single-port robotic radical prostatectomy. *Urology*. 2020;144:142–146. Available from: https://doi.org/10.1016/j.urology.2020.06.029.
2. Kaouk, et al. Extraperitoneal single-port robot-assisted radical prostatectomy: initial experience and description of technique. *BJU Int*. 2020;125:182–189. Available from: https://doi.org/10.1111/bju.14885.
3. Kaouk, et al. Single-port robot-assisted radical prostatectomy: first clinical experience using the SP surgical system. *Urology*. 2019;124:309. Available from: https://doi.org/10.1016/j.urology.2018.10.025.
4. Dobbs, et al. Single-port robot-assisted laparoscopic radical prostatectomy: initial experience and technique with the da Vinci SP platform. *BJU Int*. 2019;124:1022–1027. Available from: https://doi.org/10.1111/bju.14864.
5. Glascock, et al. Carbon dioxide homeostasis during transperitoneal or extraperitoneal laparoscopic pelvic lymphadenectomy: a real-time intraoperative comparison. *J Endourol*. 1996;10:319–323. Available from: https://doi.org/10.1089/end.1996.10.319.
6. Covas, et al. Early outcomes of single-port robot-assisted radical prostatectomy: lessons learned from the learning-curve experience. *BJU Int*. 2021;127:114–121. Available from: https://doi.org/10.1111/bju.15158.
7. Abaza, et al. Adoption of single-port robotic prostatectomy: two alternative strategies. *J Urol*. 2020;34(12):1230–1234. Available from: https://doi.org/10.1089/end.2020.0425.
8. Ng, et al. Robot-assisted single-port radical prostatectomy: a phase 1 clinical study. *Int J Urol*. 2019;26:878–883. Available from: https://doi.org/10.1111/iju.14044.
9. Kim, et al. Single-port robot-assisted radical prostatectomy with the da Vinci SP system: a single surgeon's experience. *Investig Clin Urol*. 2020;61:173–179. Available from: https://doi.org/10.4111/icu.2020.61.2.173.
10. Agarwal, et al. Initial experience with da Vinci single-port robot-assisted radical prostatectomies. *Eur Urol*. 2019;77:373–379. Available from: https://doi.org/10.1016/j.eururo.2019.04.001.
11. Kaouk, et al. Extraperitoneal vs transperitoneal single port robotic radical prostatectomy: a comparative analysis of perioperative outcomes. *J Urol*. 2020;203(6):1135–1140. Available from: https://doi.org/10.1097/JU.0000000000000700.
12. Congnard, et al. Outpatient robot-assisted radical prostatectomy: a feasibility study. *Urology*. 2019;128:16–22. Available from: https://doi.org/10.1016/j.urology.2019.01.050. Epub 2019 Mar 18.

CHAPTER

4 The re-discovery of alternative access to the pelvic fossa: the role of the single-port robotic platform

Juan Garisto[1,2], Venkat M. Ramakrishnan[1,3], Riccardo Bertolo[4] and Jihad Kaouk[5]

[1]*Department of Surgery, Brigham and Women's Hospital, Harvard Medical School, Boston, MA, United States*

[2]*Division of Urology, Veterans Affairs Medical Center, Boston, MA, United States*

[3]*Division of Urology, Brigham and Women's Hospital, Boston, MA, United States*

[4]*Department of Urology, San Carlo di Nancy Hospital, Rome, Italy*

[5]*Department of Urology, Glickman Urological and Kidney Institute, Cleveland Clinic, Cleveland, OH, United States*

4.1 Introduction

The advent of robotics in medicine has generated new avenues in surgical care. Robotic platforms provide surgeons with enhanced visualization and improved degrees of freedom of intraoperative movement and articulation. Urology has consistently been at the forefront of robotics adoption worldwide, as operating within the pelvic fossa is the quintessential application for the marriage of technology and surgery.

Before robotics was widely adopted in pelvic surgery, there was a natural, gradual transition in operative approaches, first from open to laparoscopic, then to the laparo-endoscopic single-site surgery (LESS) era, and finally to the present, with the introduction of robotic laparoscopy. LESS was initially developed with the goal of expediting postoperative recovery, reducing pain, and enhancing cosmesis.[1] In the past decade alone, all major urological procedures were implemented using LESS with subsequent improvement and evolution. Despite this, interest in LESS declined due to a steep learning curve and significant technical challenges of a smaller workspace when compared against standard laparoscopic approaches.

Single-Port Robotic Surgery in Urology. DOI: https://doi.org/10.1016/B978-0-323-91906-7.00003-X

The introduction of the da Vinci surgical robotics platform (Intuitive Surgical, Inc., Sunnyvale, California, United States) paved the way for a modern reintroduction of LESS. Kaouk and colleagues reported on their initial clinical series of robotic LESS (R-LESS) in 2009 utilizing a da Vinci Model S.[2] Despite improving upon the range of motion with suturing and dissection over traditional LESS, R-LESS surgeons continued to face limits on motion and instrument clashing in and outside the operative field.

Surgeons have always been eager to refine their technique and reduce patient morbidity. To this end, each iteration of the da Vinci robot has further reduced preceding technical limitations, enhanced surgical efficacy, and spurred the creation of new surgical techniques. It is in this vein that the da Vinci single-port (SP) platform was developed, receiving US Food and Drug Administration approval in May 2018[3,4] and facilitating the conversion of many high-morbidity inpatient surgeries into outpatient procedures[5] with equivalent outcomes and a drastically improved quality of life.

In this chapter, we discuss the application of the SP platform in the context of alternative access into the pelvic fossa, with a focus on applications targeting the prostate, bladder, and ureters (Table 4.1).

4.2 The evolution of the robotic surgical approaches to the pelvic fossa

The da Vinci surgical robotics platform is the most widely used among the robotic platforms available on the market. It was made commercially available in 1999 as a 3-arms system that rivaled the traditional laparoscopic approaches of the time.[6] After modifying the original platform in 2003 with adding a fourth arm, the da Vinci Model S. augmented the surgical experience in 2006 with high-definition optics (720 px) and a simplified set-up process that facilitated the surgical access. In 2009, the Si further enhanced the quality of the intraoperative visualization with even higher-definition optics (1080 px) and offered the integrated support of fluorescence by the availability of infra-red cameras, introduced more complex instrumentation, and improved educational access to surgeons, fellows, and residents with the potential for a dual-console support, networking, and an accompanying surgical skills simulator. The Xi capitalized on this growth with 3D-optics, the ability to dock the side-cart alongside a supinated patient via a rotating boom, and offered uniform port sizes to further optimize the camera and the port placement.[7]

In 2018, single port surgery emerged as the next advance in robotic laparoscopy via the SP platform. The SP utilizes a single, multichannel port from which three working multijointed instruments and a high-definition camera emerge. From a surgeon's perspective, single-port surgery further broadens the operative horizon by reducing the number of required ports, reducing the risk of incisional hernias, extending surgical access to more areas of the body (such as narrow-access

Table 4.1 Single-port robotic surgical approaches into the pelvic fossa.

Access	Procedures studied	Port placement	Comments
Transperitoneal	Radical and simple prostatectomy Radical cystoprostatectomy & pelvic exenteration Partial cystectomy Ureteral reimplantation Kidney transplantation	2 cm above umbilicus 5 cm above umbilicus 2 cm above umbilicus 5 cm above umbilicus Periumbilical incision	- Wide operative space - Trendelenburg position - Does not require lithotomy positioning unless access to vagina (for pelvic exenteration) is desired
Extraperitoneal	Radical prostatectomy	2 cm below umbilicus	- Supine position - Does not require Trendelenburg position - Limited pelvic lymph node dissection - Avoids intraperitoneal insufflation
Perineal	Perineal prostatectomy	Perineal semilunar	- Lithotomy position - Does not require Trendelenburg position - Useful in those with complex abdominal surgical history or restrictive lung disease
Transvesical	Radical and simple prostatectomy	Suprapubic	- Limited pelvic lymph node dissection - Facilitates concurrent cystolitholapaxy - Pneumovesicum limits systemic CO_2 absorption
Retzius-sparing	Radical prostatectomy	2 cm above umbilicus	

cm, *centimeters.*

transoral, transanal, perineal, and extraperitoneal applications). Moreover, it can potentially be employed in an even younger age group with reduced morbidity.

4.3 Surgical approaches to the pelvic fossa using the da Vinci SP robot

Fig. 4.1

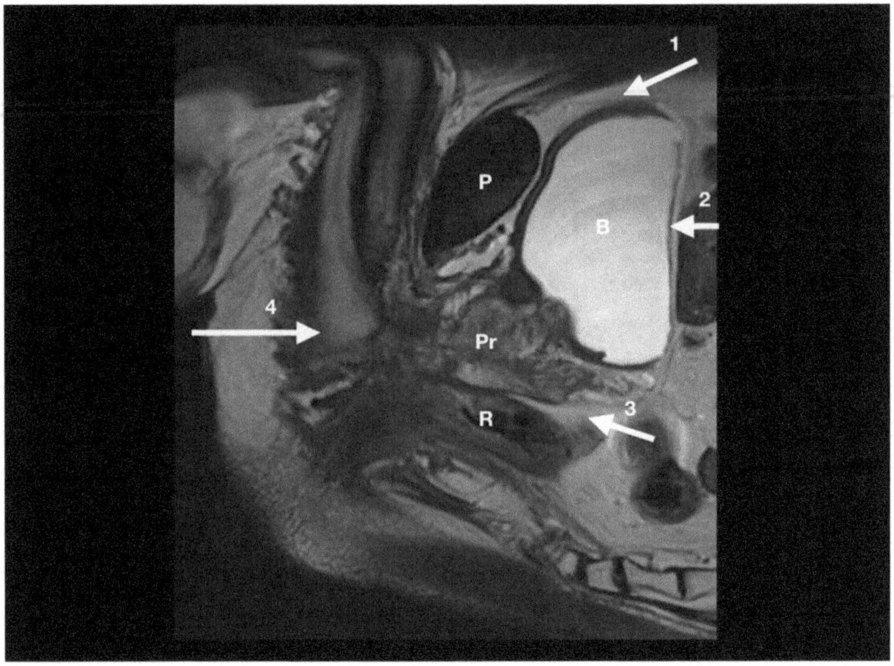

FIGURE 4.1

Pelvic MRI in a sagittal view. Each arrow demonstrates possible access into the pelvic fossa. (1) Transperitoneal & Extraperitoneal, (2) Transvesical, (3) Retzius-sparing, and (4) Perineal.

4.3.1 Transperitoneal approach

The transperitoneal approach to the pelvic fossa is the standard and the most widely used one among all the "urological" approaches (Fig. 4.2).

Via a transperitoneal approach to gain the access to the pelvic fossa, the SP platform has already been employed to perform a variety of major urological surgery, including radical prostatectomy,[8] radical cystoprostatectomy[9] with or without intracorporeal urinary diversion, pelvic exenteration, ureteral reimplantation,[10] and kidney transplant.[11] The principles of the surgical approaches for each of these cases are similar and described as follows. A small 2.5 cm incision is first made above the umbilicus, followed by an additional 5–8 mm incision at the level of the lateral abdominal wall if an assistant port is desired. Once the peritoneal access and the pneumoperitoneum are established, the SP can be used to mobilize the bladder (Fig. 4.3), remove the periprostatic fat, and expose and incise the endopelvic fascia in the traditional fashion.

For radical prostatectomy, the dorsal venous complex (the Santorini plexus) can be identified and managed before the bladder neck is incised; alternatively,

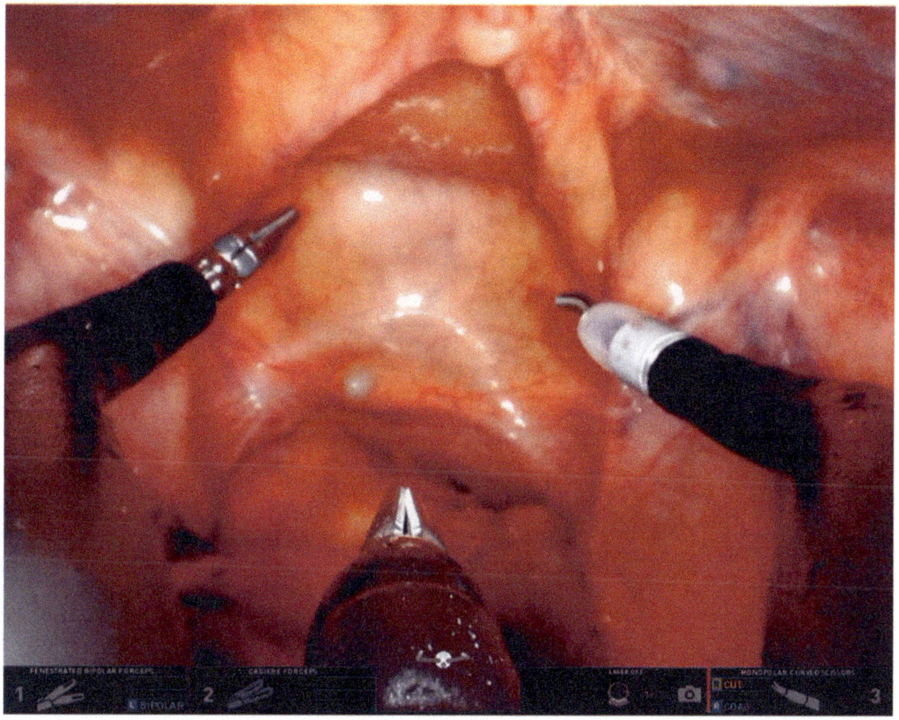

FIGURE 4.2

Intraoperative view after gaining the transperitoneal access to perform single-port robotic prostatectomy.

a selective ligation of venous sinuses of the plexus can be performed once the apical dissection has been performed. The seminal vesicles and the vas deferens are delivered and the latter is ligated. The prostate is then ligated circumferentially prior to the apical and the urethral dissection, to complete prostatectomy. Finally, the vesico-urethral anastomosis is performed to restore the integrity of the urinary system. This approach provides the most (visually) complete surgical exposure of the pelvic fossa among the discussed approaches. Further, it facilitates the broadest access to the pelvic lymph nodes, allowing for an extended and super-extended lymph-nodes dissection.

The earliest uses of the SP employed the robotic platform in radical prostatectomies via a transperitoneal approach.[12] The first report included two cases that were successfully completed without converting to a traditional robotic or open approach, with an operative time of 40 minutes and without perioperative complications. The approach has also been described in the context of distal ureteral repairs. Kaouk and colleagues discussed the use of the SP in three patients with distal benign ureteral strictures and reported no intraoperative complications.[10]

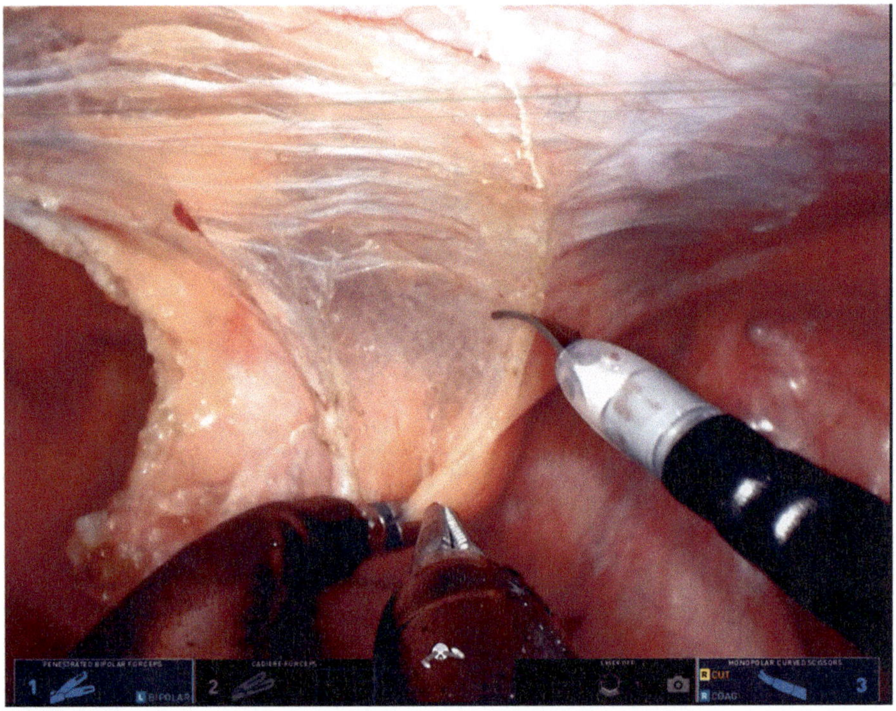

FIGURE 4.3

Incision of the peritoneum to drop down the bladder during transperitoneal single-port robotic prostatectomy.

The wide maneuverability of the SP platform provided an adequate range of motion to reach the pelvic fossa without hampering the ability to perform robot-assisted reconstructive surgery.

Turning our focus to the bladder, radical cystoprostatectomy or pelvic exenteration remains the gold standard treatment for muscle-invasive and refractory noninvasive bladder cancers. There was an exponential increase in the use of robot-assisted approaches to radical cystoprostatectomy between 2002 and 2012, starting at 0.7% of all cystectomies to 18.5% in that time. Zhang and colleagues were one of the first to use the SP in radical cystoprostatectomy applications.[9,13] He described a series of four patients who underwent SP robot-assisted radical cystoprostatectomy with an operative time of 270 minutes and EBL of 250 cc. No intraoperative complications were reported, and all procedures were completed successfully.

Finally, precedent has also been established for use of the platform in robotic-assisted kidney transplantation. Six cases were completed successfully without conversion to an alternative approach, and two of these cases utilized transperitoneal access. With the transperitoneal approach, the cecum and the right colon were medialized. With an extraperitoneal approach (utilized for the remaining four cases and discussed in further detail below), direct access to external iliac vessels was possible. In both approaches, the external iliac artery and vein were dissected. The right anterolateral aspect of the bladder was then prepared for ureteroneocystostomy.[11]

4.3.2 Extraperitoneal approach

Using the SP via an extraperitoneal approach is particularly useful in patients with a prior history of major abdominal surgeries. Moreover, as with the transvesical and the transperineal (discussed in section 4.3.3. and 4.3.5., respectively), the extraperitoneal approach avoids the pitfalls of the pneumoperitoneum and the bowel-related complications (reduced bowel mobilization, decreased likelihood of encountering or causing adhesions, and reduced rates of postoperative ileus).[14] Conversely, the decision to pursue a trans- or an extra-peritoneal approach has consistently leaned towards the former, namely due to the wider working space and the notorious anatomical landmarks.

With the extraperitoneal approach to the pelvis, the prostate is accessed via the anterior rectus fascia. A 2–3 cm incision is made about one fingerbreadth below the umbilicus (Fig. 4.4).

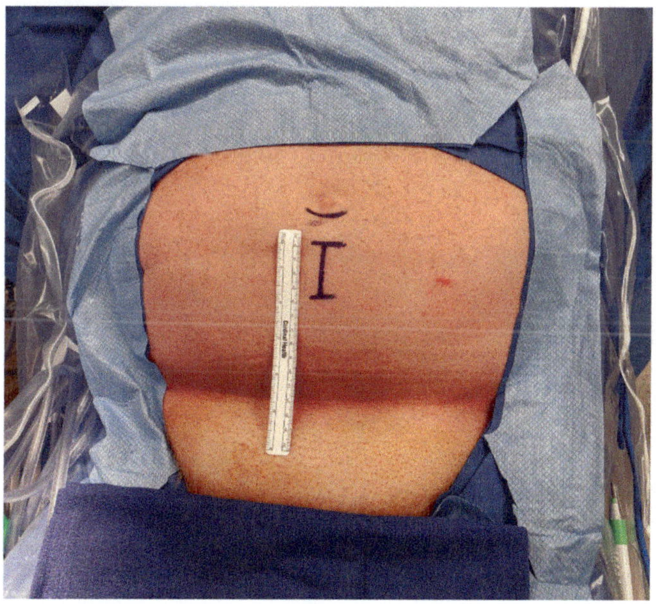

FIGURE 4.4

The 2–3 cm incision made one fingerbreadth below the umbilicus to gain the pelvis via an extraperitoneal approach.

The anterior rectus sheath is incised sharply and dissected. A finger is then used to dissect the preperitoneal space prior to the placement of a SPACEMAKER surgical balloon dissector system (Covidien, Dublin, Ireland). The working area in the Space of Retzius is further maximized by inflating the balloon with 300–400 cc of air under direct visualization. A GelPOINT wound protector and retraction system (Applied Medical, Rancho Santa Margarita, California, United States) is placed into the operative field and secured under the rectus fascia. This is followed by the insertion of a 25 mm SP cannula via the GelPOINT GelSeal cap (Figs. 4.5 and 4.6).

An additional 12-mm laparoscopic port can be inserted through the cap or into the right pelvic fossa to serve as an assistant port. The prostatectomy then proceeds in the usual fashion, by following the standard surgical steps (Figs. 4.7, 4.8, and 4.9).

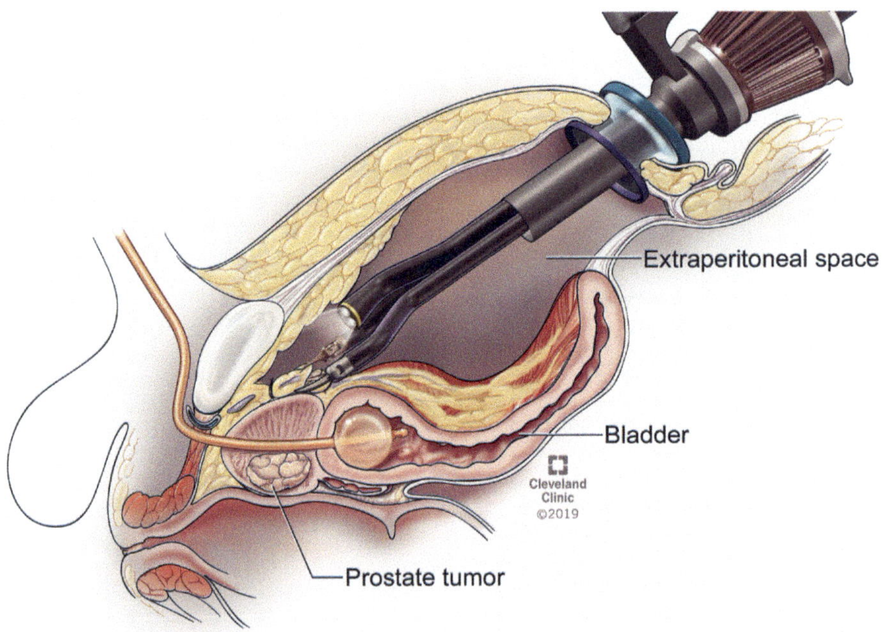

FIGURE 4.5

The cartoon summarizes the extraperitoneal access to perform single-port prostatectomy. The GelPOINT wound protector and retraction system (Applied Medical, Rancho Santa Margarita, California, United States) is placed to gain the direct access to the extraperitoneal space. It accommodates the 25-mm multichannel SP cannula. An instrument arm, which is attached to the patient side cart, contains four instrument drives that control the 12 × 10-mm articulating robotic camera and three 6-mm double-jointed articulating robotic instruments.

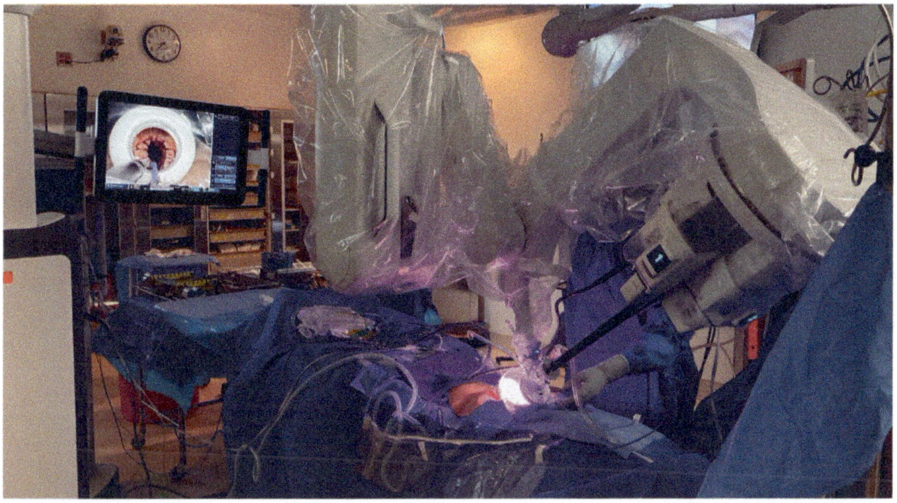

FIGURE 4.6

The SP surgical platform docked for extraperitoneal single-port robotic prostatectomy.

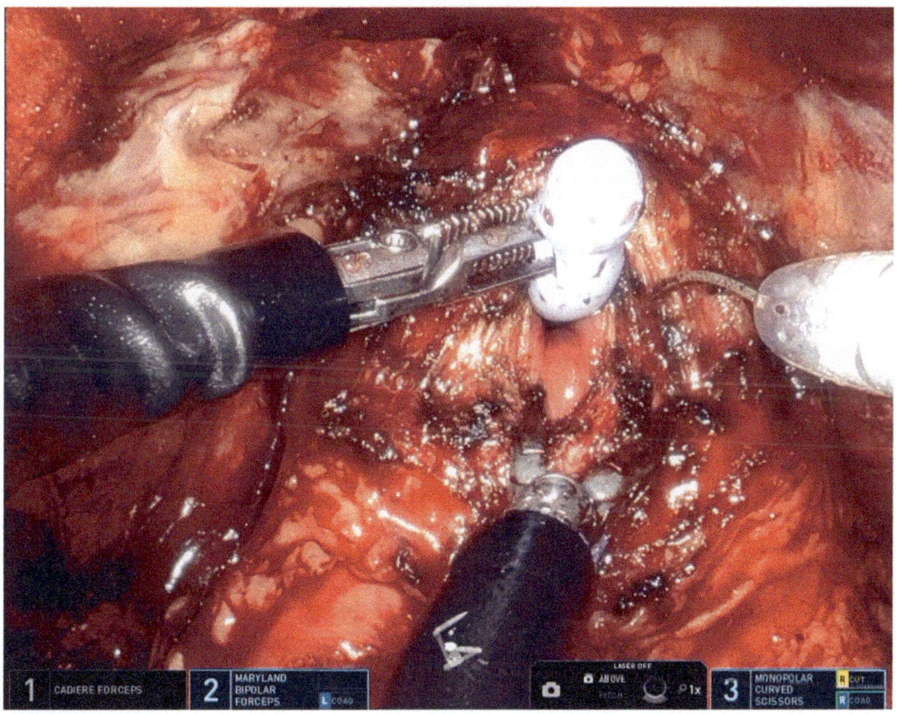

FIGURE 4.7

Bladder neck dissection during extraperitoneal single-port robotic prostatectomy.

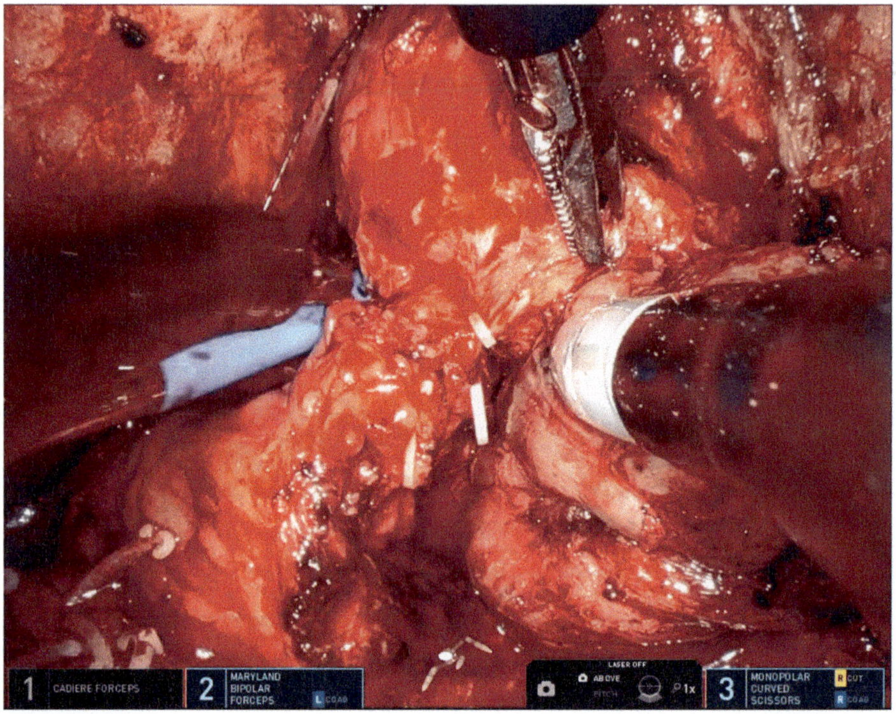

FIGURE 4.8

Dissection of the neuro-vascular bundles during extraperitoneal single-port nerve-sparing robotic prostatectomy.

The first use of the SP in extravesical prostate surgery included 10 patients diagnosed with localized disease. The series reported no conversions or intraoperative complications, with a mean operative time of 197 minutes and EBL ranging from 50 to 400 cc. This early experience conveyed favorable outcomes and possible advantages, such as the ability to avoid a steep Trendelemburg positioning (particularly beneficial in obese patients and those with chronic obstructive pulmonary disease),[15] minimal postoperative pain, and the potential for a same-day discharge.[5] One major disadvantage of this technique is that by the nature of the approach, extended lymph node dissections cannot be performed. In a recent comparative study analyzing the perioperative outcomes of extraperitoneal versus the standard transperitoneal one, the authors reported a shorter hospitalization and a decreased need for postoperative narcotics favoring the extraperitoneal approach to SP-RARP.[16]

4.3.3 Transvesical approach

Previously employed in preclinical models for partial prostatectomy, the transvesical approach can now also be used via the SP platform to perform robotic

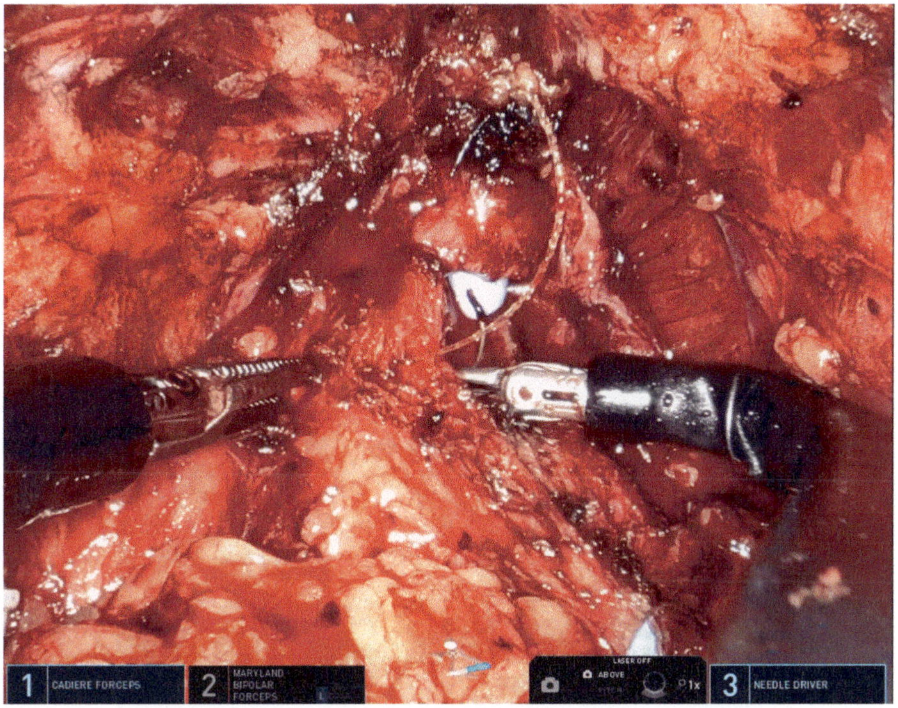

FIGURE 4.9

Vesico-urethral anastomosis performed during extraperitoneal single-port robotic prostatectomy.

radical[17,18] and simple[19] prostatectomies. Positioning and port placement are the same for both procedures. The patient is placed in a supine position and an incision is made 4 cm above the pubic symphysis (Fig. 4.10).

After the bladder is distended with irrigant via a urethral catheter, a GelPOINT trocar is percutaneously inserted into the bladder followed by evacuation of the irrigant. Once the bladder is empty, a multichannel SP cannula is inserted through the GelPOINT GelSeal cap. The pneumovesicum is established with carbon dioxide insufflation, and a domain is established that allows for the accommodation of the articulating 3D robotic camera and the three 6-mm double-jointed instruments (Figs. 4.11 and 4.12).

After the robot is docked, the intravesical view is focused on the bladder trigone (Fig. 4.13).

For radical prostatectomy, the gland is accessed by incising the bladder neck distal to the trigone and ureteral orifices, facilitating immediate visualization of the peripheral zone of the prostate (Fig. 4.14).

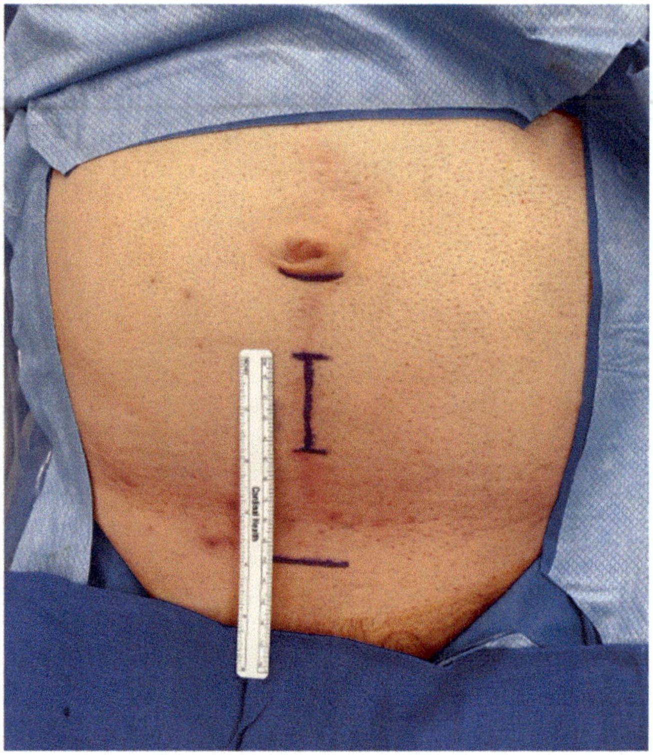

FIGURE 4.10

The 4 cm incision made above the pubic symphysis to gain the prostate via a transvesical approach.

Posterior dissection proceeds inferiorly toward the prostatic apex. The seminal vesicles and vas deferens are both mobilized (Fig. 4.15), and the infratrigonal intravesical incision is then extended circumferentially to 12 o'clock (adjacent to, but not involving, the dorsal venous complex) to complete the bladder neck dissection and subsequent anterior dissection of the gland (Fig. 4.16).

Finally, the anastomosis is completed (Figs. 4.17, 4.18, and 4.19).

The transvesical approach has several potential benefits and one key consideration. Firstly, as with the extravesical approach, it avoids the risks of intraperitoneal access and can be used in those patients with a history of prior abdominal surgeries. Secondly, it avoids the need for a steep Trendelenburg position. Thirdly, because the pneumovesicum only is induced (not the pneumoperitoneum), little, if any, of the carbon dioxide insufflation is absorbed and does not have the traditional effects on operative physiology. Perhaps one

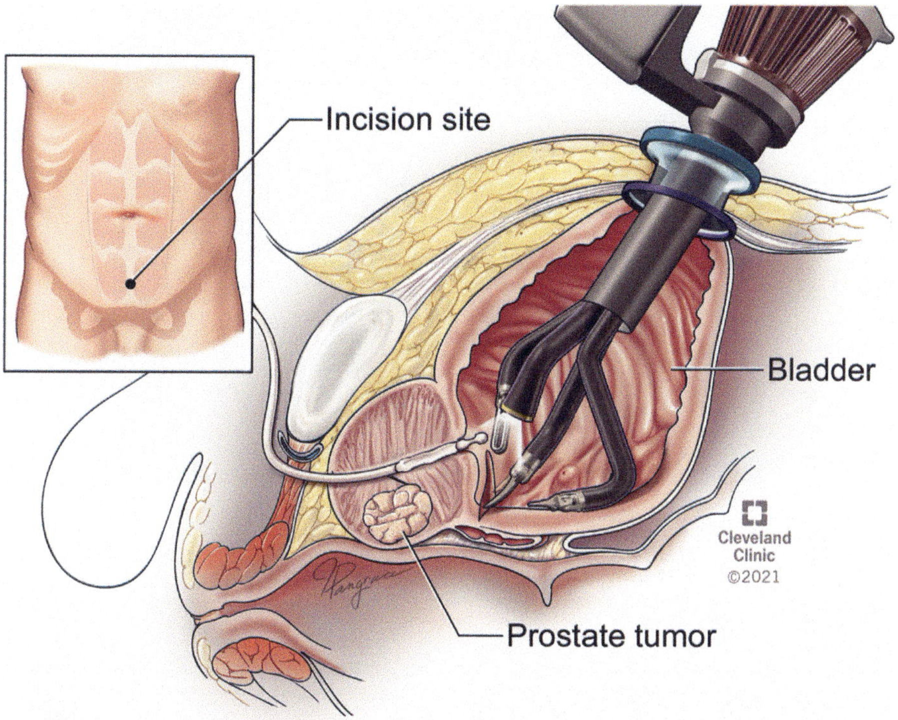

FIGURE 4.11

The cartoon summarizes the transvesical access to perform single-port prostatectomy. After the bladder is distended with irrigant via a urethral catheter, a GelPOINT trocar is percutaneously inserted into the bladder followed by evacuation of the irrigant. Once the bladder is empty, a multichannel SP cannula is inserted through the GelPOINT GelSeal cap. The pneumovesicum is established with carbon dioxide insufflation, and a domain is established that allows for the accommodation of the articulating 3D robotic camera and the three 6-mm double-jointed instruments.

of the biggest benefits relative to the transvesical approach is that patients with significant cardiopulmonary comorbidities may be able to avoid general anesthesia and instead receive spinal anesthesia via an epidural. What remains unknown is whether such an approach can be utilized in patients with large prostate glands or in patients with preexisting significant lower urinary tract symptoms. In patients with high-volume bladders, trabeculation, and diverticula, the transvesical approach might not be ideal. Preoperative imaging of the lower urinary tract and cystoscopy for direct evaluation could be beneficial in such circumstances.

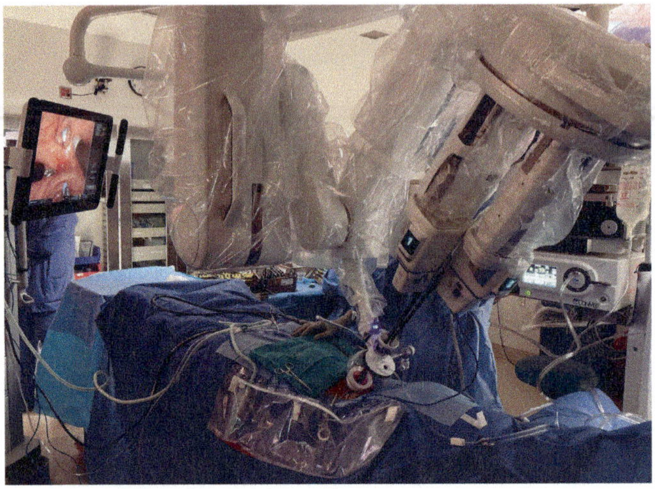

FIGURE 4.12

The SP surgical platform docked for transvesical single-port robotic prostatectomy.

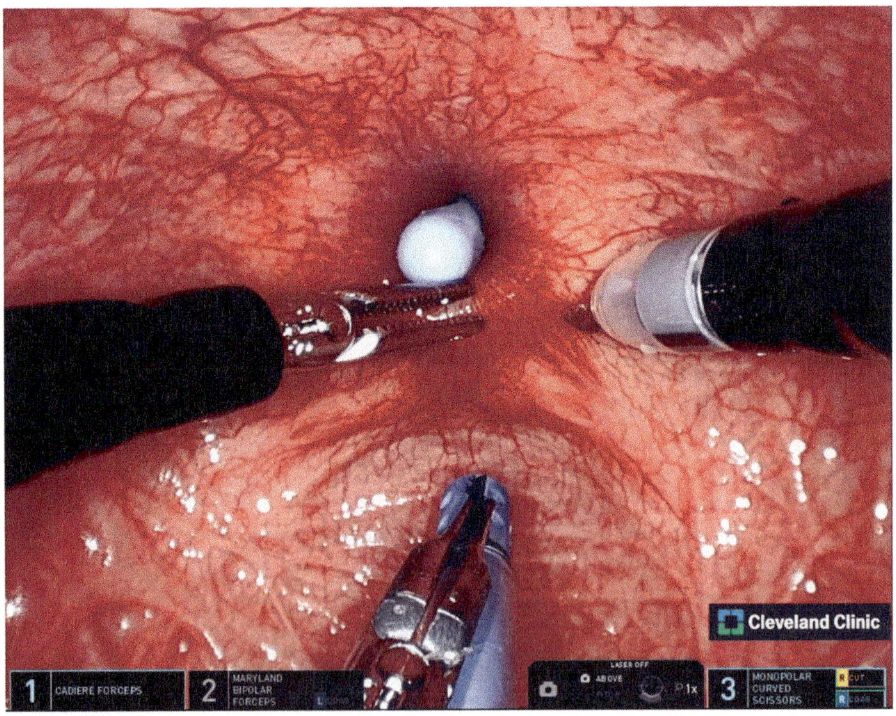

FIGURE 4.13

Intravesical view focused on the bladder trigone during transvesical single-port robotic prostatectomy. The bladder neck and the ureteral orifices are evident.

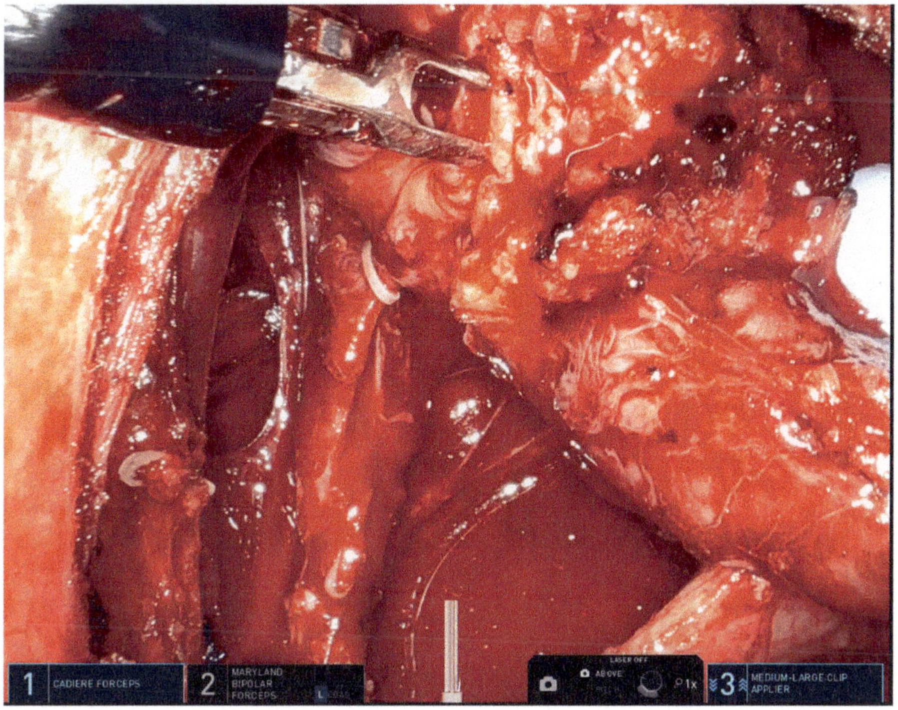

FIGURE 4.14

Dissection of the neuro-vascular bundles during transvesical single-port nerve-sparing robotic prostatectomy.

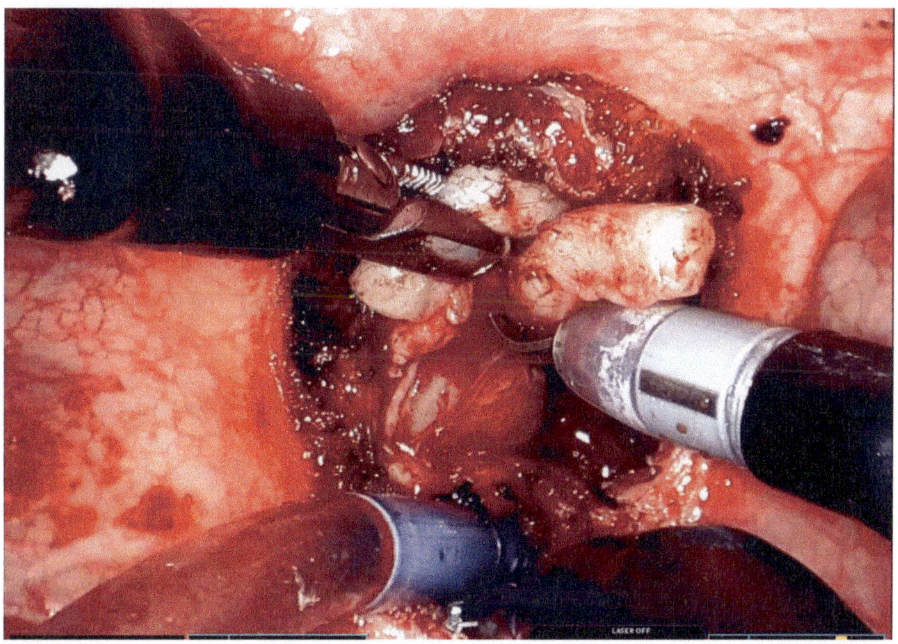

FIGURE 4.15

Dissection of the seminal vesicles during transvesical single-port robotic prostatectomy.

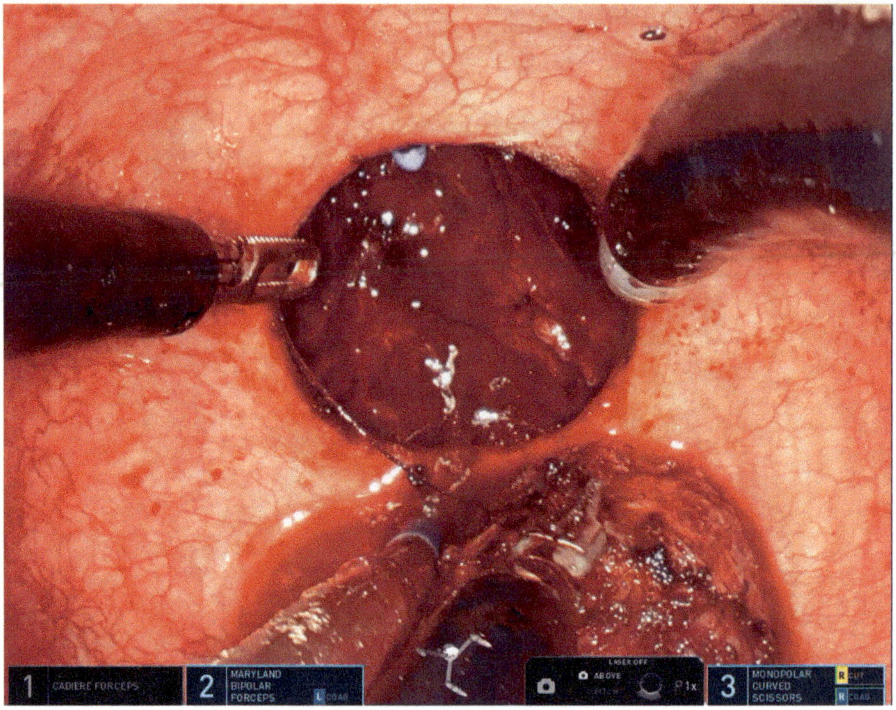

FIGURE 4.16

Intravesical view of the fully excised bladder neck after transvesical single-port robotic prostatectomy is completed (the specimen is put into the bladder lumen).

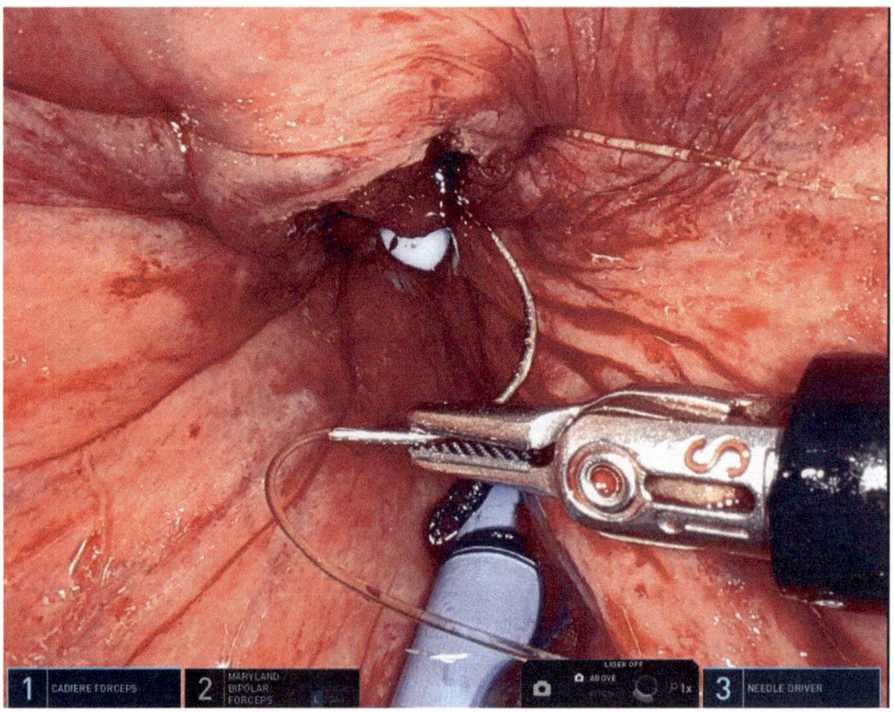

FIGURE 4.17

Vesico-urethral anastomosis performed during transvesical single-port robotic prostatectomy.

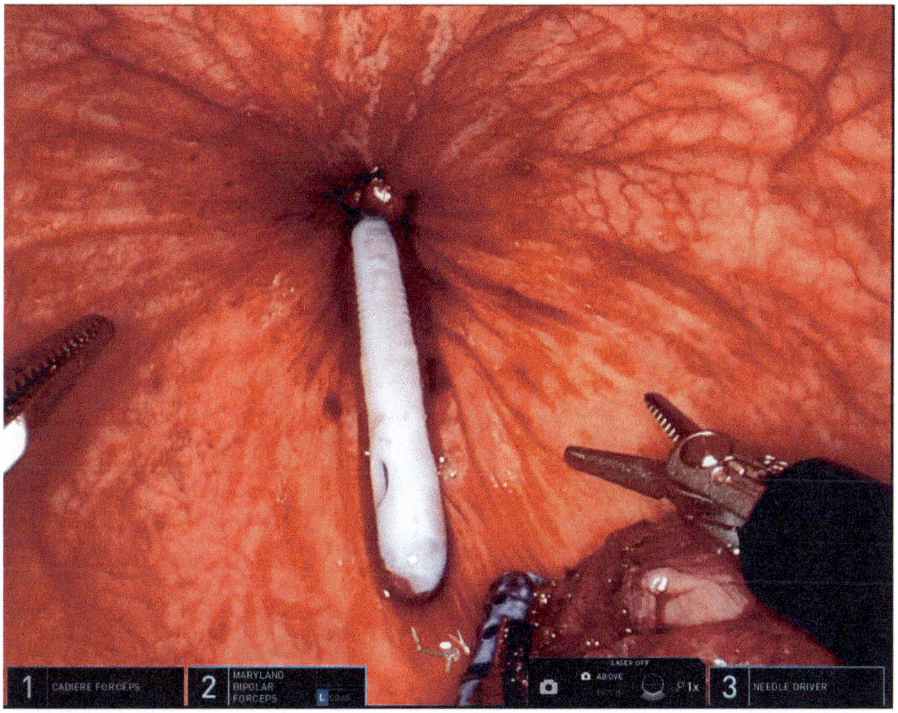

FIGURE 4.18

Completed vesico-urethral anastomosis performed after transvesical single-port robotic prostatectomy.

4.3.4 Retzius-sparing approach

A Retzius-sparing (RS) approach for single-port prostatectomy is one option for surgeons who wish to achieve a smaller surgical dissection while conferring a potentially earlier return of urinary continence. The RS robot-assisted radical prostatectomy was first described by Galfano et al.[20] using a multiarm robot. Broadly, the surgery represents the continuation in an anterior direction of the posterior approach to the vasa and seminal vesicles through a posterior peritoneal dissection (Fig. 4.20).

The gland is then dissected inferiorly off of Denonvillier's fascia and the bladder neck is identified posteriorly and incised. Temporary sutures are often placed at the bladder neck to facilitate its eventual closure. Afterwards, the prostate is bluntly dissected off the dorsal venous complex before the apex of the gland is isolated and incised to complete the prostatectomy (Figs. 4.21 and 4.22).

Upon completing the bladder neck anastomosis (Fig. 4.23), the parietal peritoneum is then closed at the Space of Douglas.

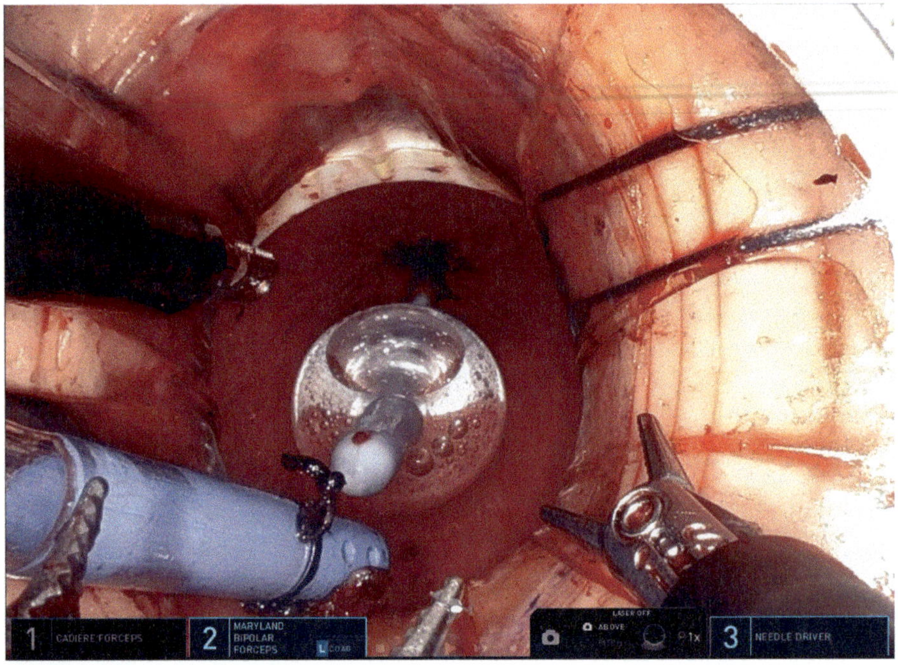

FIGURE 4.19

Intraoperative view after completed vesico-urethral anastomosis performed after transvesical single-port robotic prostatectomy. The catether balloon has been inflated. The robotic camera is retracted at the level of the Alexis wound protector-retractor of the GelPOINT advanced access platform.

The SP RS radical prostatectomy was first described in a preclinical model demonstrating the suitability of the procedure.[21] This initiative was followed by seven SP RS cases, with mean operative times of less than 200 minutes confirming feasibility of the approach.[22]

Globally, the SP RS radical prostatectomy has several advantages and disadvantages. Firstly, it avoids the pubo-urethral ligaments and complications resulting from manipulation of the dorsal venous complex of Santorini. Because of this and a smaller surgical dissection, patients undergoing RS radical prostatectomy have the potential for an earlier return of urinary continence postoperatively. The avoided "violation" of the Retzius space has been demonstrated by Dr. Bassett, who presented during the June 2021 Society of Urologic Robotic Surgeons "SP marathon" an impressive synopsis between the space of Retzius before starting and after completing a Retzius-sparing robotic prostatectomy (Fig. 4.24).

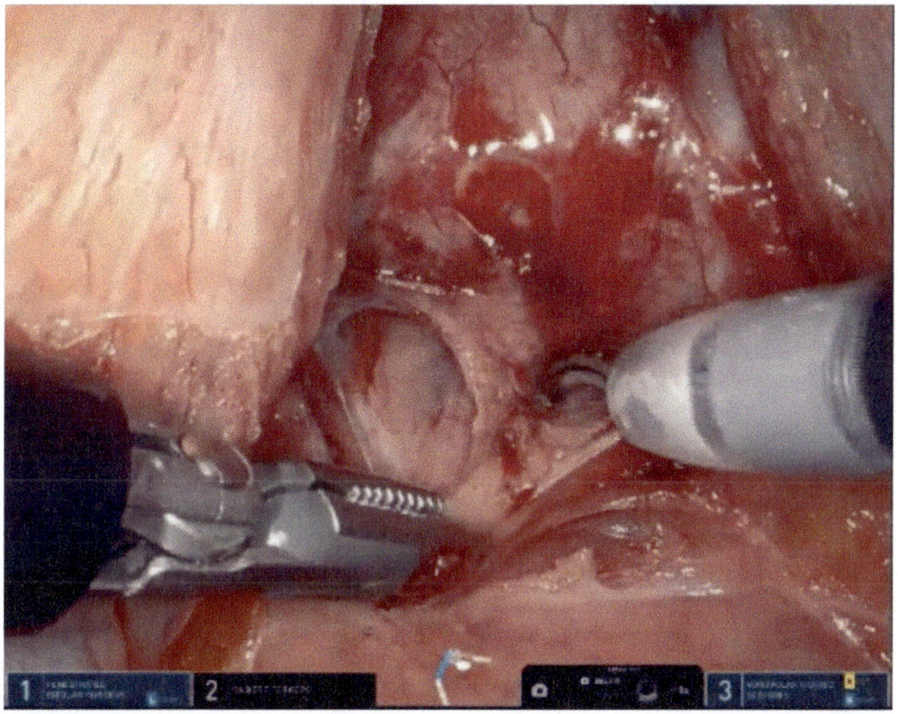

FIGURE 4.20

The incision of the peritoneum typical of the posterior approach during Retzius-sparing single-port robotic prostatectomy.

However, the approach has a separate learning curve and potential complications. Moreover, the approach confers a smaller working space, which makes operating on larger glands technically challenging according to several reports. Other challenges include no lateral aiming point when dissecting the lateral pedicles of the prostate, an inability to visually assess the bladder neck division and verify the position of the ureteric orifices. Finally, the surgeon has to deal with an inverted relationship between the bladder and the prostate during dissection and reconstruction.

4.3.5 Transperineal approach

Transperineal access is one of the oldest approaches in addressing disease of the pelvic fossa, relying on a small perineal incision. Young first described it when he performed an open radical perineal prostatectomy in 1905.[23]

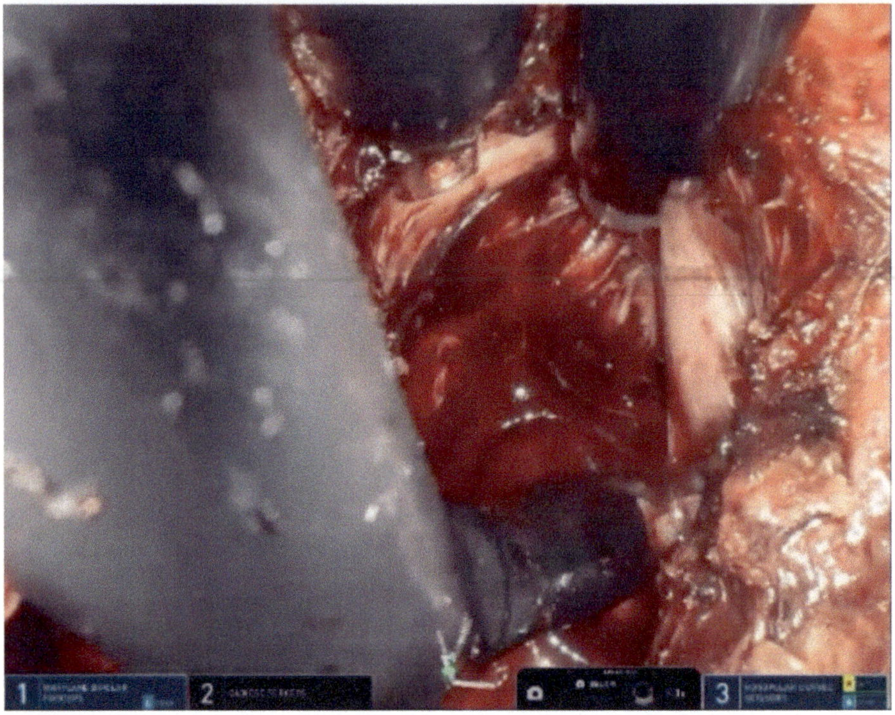

FIGURE 4.21

Progressive dissection toward the prostate apex during Retzius-sparing single-port robotic prostatectomy.

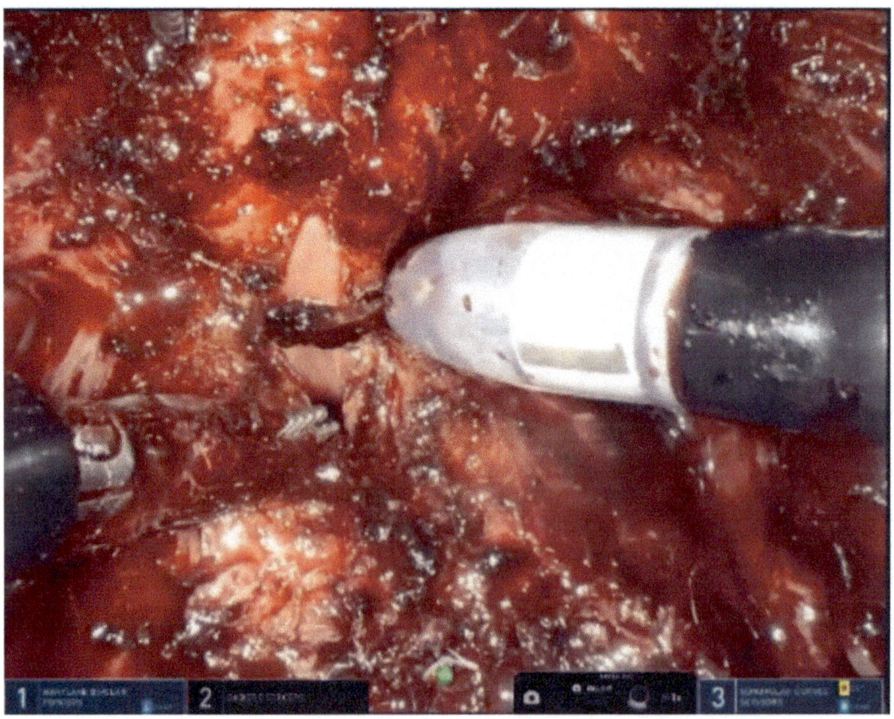

FIGURE 4.22

Section of the posterior aspect of the urethral stump during Retzius-sparing single-port robotic prostatectomy.

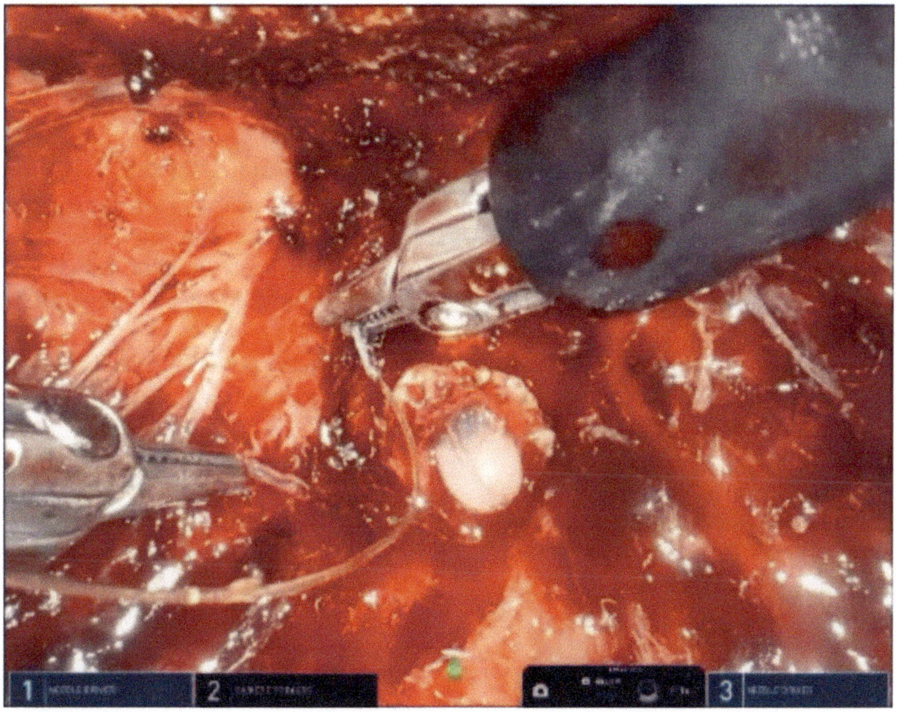

FIGURE 4.23

Vesico-urethral anastomosis performed during Retzius-sparing single-port robotic prostatectomy.

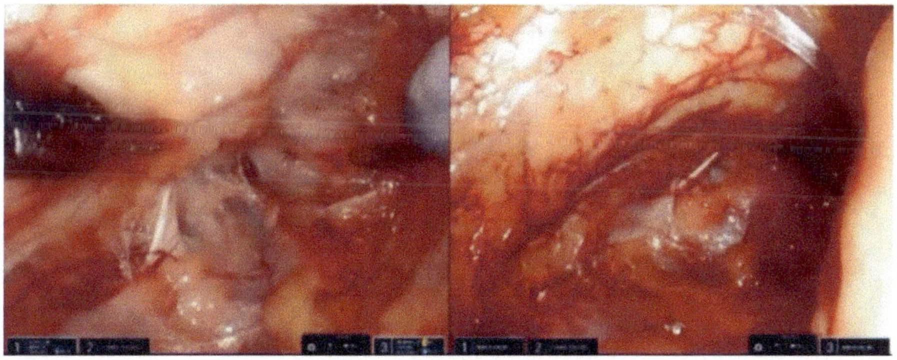

FIGURE 4.24

The space of Retzius before starting (right) and after completing (left) a Retzius-sparing robotic prostatectomy. The avoided "violation" of the Retzius space is clearly demonstrated. Presented by Dr. Bassett during the June 2021 Society of Urologic Robotic Surgeons (SURS) "SP marathon."

However, with the development of new routes—the retropubic approach for radical prostatectomy, for instance, the perineal approach became less favored because the anatomy was more complex from below. In addition, the high position of the prostate in the pelvis and the surgeon's narrow operative view were further challenges that hampered broad use of the approach.

Laydner and colleagues first investigated the applicability of the Si platform in robotic transperineal prostatectomy,[24] evaluating the feasibility of transperineal access in preclinical, cadaveric models. Ultimately, the use of a multiarm robotic system was technically challenging and time consuming,[25] both of which were largely rectified with the SP. The technique has typical steps, unique of the approach (Figs. 4.25 and 4.26).

It is reported in detail in the dedicated chapter of the book. In recent days, the advent of the SP platform of the market has prompted the resurgence of robotic perineal prostatectomy.[26]

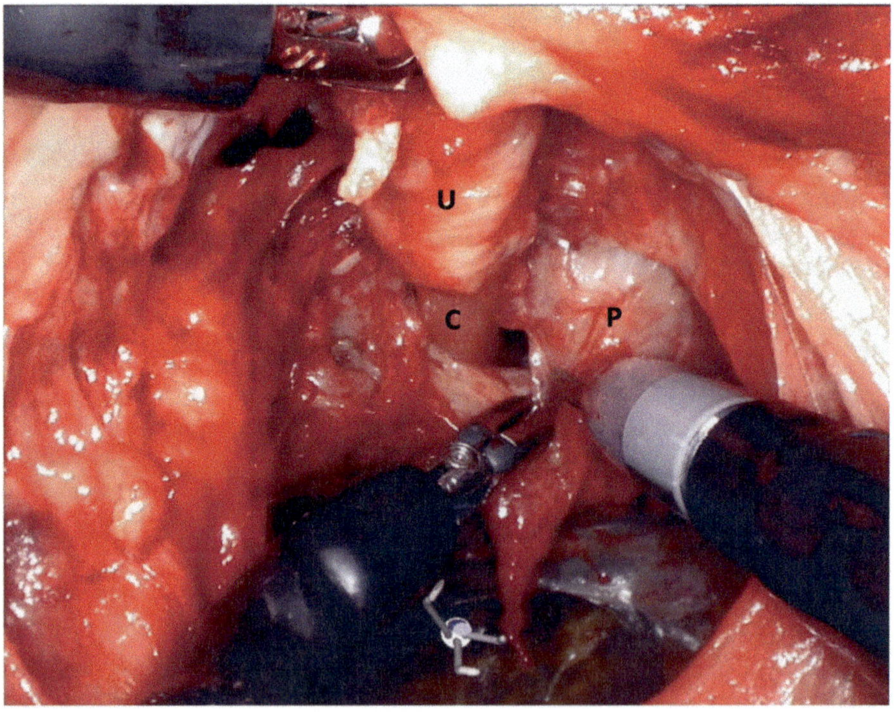

FIGURE 4.25

Transperineal robot-assisted radical prostatectomy performed on a preclinical model by using the da Vinci SP robotic platform: the prostate apex dissected along with the urethra is shown (P: prostate; U: urethra; C: catheter).

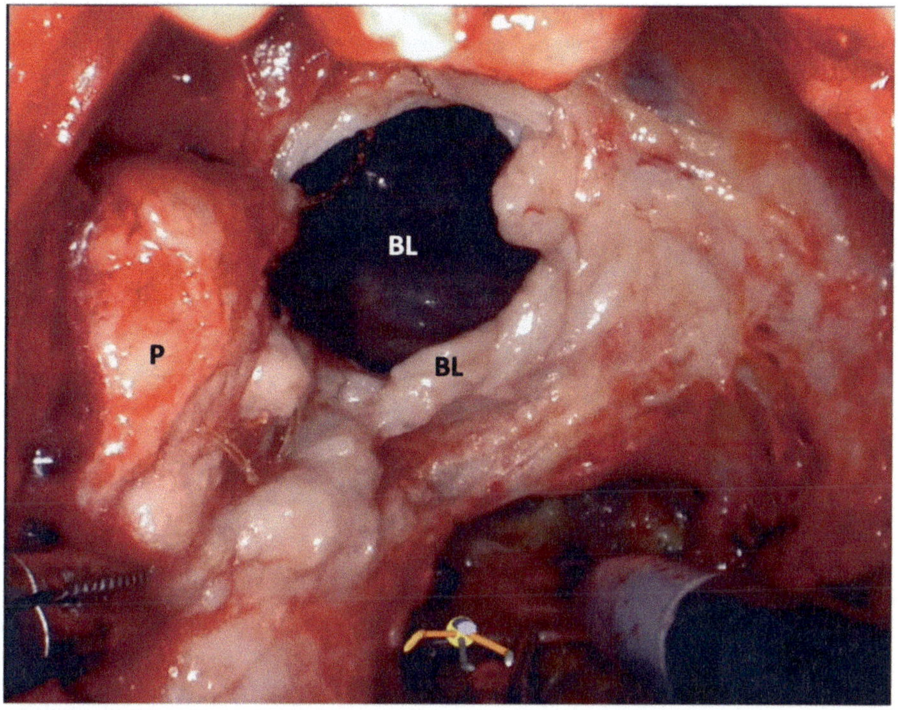

FIGURE 4.26

Bladder neck dissection during transperineal robot-assisted radical prostatectomy performed on a preclinical model by using the da Vinci SP robotic platform (P: prostate; BL: bladder lumen).

With the typical drawbacks of such approach that overcame, thanks to the SP platform, the perineal access to the pelvic fossa virtually eliminates the risks of injury to the small bowel or major vessels during the port placement. In addition, it facilitates avoidance of extensive adhesions in patients with previous abdominal wall repairs or multiple surgeries. It also offers a viable approach for obese patients and a way to obviate excessive airway and pneumoperitoneal pressures in this population.[27] Although perineal robotic prostatectomy uses carbon dioxide insufflation to improve visualization, it eliminates the need for Trendelemburg positioning and pneumoperitoneum, avoiding the associated potential complications.

4.4 Conclusions

The commentary above is not a departure from the various approaches to prostatectomy, but rather a reflection of the broad capabilities of the da Vinci SP platform and its ability to perform the surgery under various conditions with relative ease. It has the capacity to renew previously abandoned approaches due to prior issues with technical complexity and poor visualization while potentially offering oncologically equivalent outcomes compared to the traditional transperitoneal approach. In an era in which robotic prostatectomy is (i) standardized and (ii) gradually shifting to same-day surgery in many practices, the ability of the SP to offer a dramatic reduction in port number, operate in various surgical approaches, and potentially further reduce in pain and analgesic requirement make it an especially attractive tool for the global urologic armamentarium.

References

1. Liatsikos E, Kallidonis P, Kyriazis I, Al-Aown A. Stolzenburg. *J Nat Rev Urol.* 2009;6(12):654–659.
2. Kaouk JH, Goel RK, Haber GP. Robotic single-port transumbilical surgery in humans: initial report. *BJU Int.* 2009;103:366–369.
3. Bertolo R, Garisto J, Kaouk J. Novel system for robotic single-port surgery: feasibility and state of the arte in urology. *Eur Urol Focus.* 2018;4(5):669–673.
4. Kaouk J, Garisto J, Bertolo R. Robotic urologic surgical interventions performed with the single-port dedicated platform: first clinical investigations. *Eur Urol.* 2019;75(4): 684–691.
5. Wilson CA, Aminsharifi A, Sawcyn G, et al. Outpatient extraperitoneal single-port robotic radical prostatectomy. *Urology.* 2020;144:142–146.
6. Rassweiler JJ, Autorino R, Klein J, et al. *BJU Int.* 2017;120(6):822–841.
7. Van der Schans EM, Hiep MAJ, Consten ECJ, Broeders IAMJ. From da Vinci Si to da Vinci Xi: realitstic times in draping and docking the robot. *J Robot Surg.* 2020;14(6): 835–839.
8. Kaouk J, Aminsharifi A, Sawczyn G, et al. Single-port robotic urological surgery using purpose-built single-port surgical system: single-institutional experience with the first 100 cases. *Urology.* 2020;140:77–84.
9. Kaouk J, Garisto J, Elemamy M, Bertolo R. Step-by-Step technique for single-port robot-assisted radical cystectomy and pelvic lymphs nodes dissection using the da Vinci SP surgical system. *BJU Int.* 2019. Available from: https://doi.org/10.1111/bju.14744.
10. Kaouk JH, Garisto J, Eltemamy M, Bertolo R. Robot-assisted surgery for benign distal ureteral strictures: step-by-step technique using the SP surgical system. *BJU Int.* 2019;123(4):733–739.
11. Kaouk J, Eltemamy M, Aminsharifi A, et al. Initial experience with single-port robotic-assisted kidney transplantation and autotransplantation. *Eur Urol.* 2021;80(3): 366–373.
12. Kaouk J, Bertolo R, Eltemamy M, Garisto J. Single-port robot assisted radical prostatectomy: first clinical experience using the SP surgical system. *Urology.* 2019;124:309.
13. Zhang M, Thomas D, Salama G, Ahmed M. Single port robotic radical cystectomy with intracorporeal urinary diversion: a case series and review. *Transl Andro Urol.* 2020;9(2):925–930.

14. Aminsharifi A, Sawczyn G, Wilson CA, Garisto J, Kaouk J. Technical advancements in robotic prostatectomy: single-port extraperitoneal robotic-assisted radical prostatectomy and single-port transperineal robotic assisted radical prostatectomy. *Transl Androl Urol*. 2020;9(2):848–855.
15. Boczko J, Madeb R, Golijanin D, et al. Overcoming the learning curve for robot assisted laparoscopic radical prostatectomy. *Urol Clin North Am*. 2010;37:37–47.
16. Kaouk J, Aminsharifi A, Wilson CA, et al. Extraperitoneal vs transperitoneal single-port robotic radical prostatectomy: a comparative analysis of perioperative outcomes. *J Urol*. 2019;203(6):1135–1140. Available from: https://doi.org/10.1097/JU.0000000000000700.
17. Kaouk J, Sagalovich D, Garisto J. Robotic transvesical partial prostatectomy using a purpose-built single port robotic system. *BJU Int*. 2018;122(3):520–524.
18. Kaouk J, Beksac AT, Zeinab MA, Duncan A, Schwen ZR, Eltemamy M. Single-port transvesical robotic radical prostatectomy: initial clinical experience and description of technique. *Urology*. 2021;S0090–4295(21):00430–00431.
19. Kaouk J, Sawczyn G, Wilson C, et al. Single-port percutaneous transvesical simple prostatectomy using the SP robotic system: initial clinical experience. *Urology*. 2020;141:173–177.
20. Galfano A, Tripani DD, Sozzi F, et al. Beyond the learning curve of the Retzius-sparing approach for robot-assisted laparoscopic radical prostatectomy: oncologic and functional results of the first 200 patients with >1 year of follow-up. *Eur Urol*. 2013;64(6):974–980.
21. Ng C-F, Chan ES, Teoh JC. The use of the da Vinci SP System for Retzius-sparing radical prostatectomy in cadaveric Model. *Urology*. 2019;125:260.
22. Bassett JC, Salibian S, Crivellaro S. Single-port retzius-sparing robot-assisted radical prostatectomy: feasibility and early outcomes. *J Endourol*. 2022. Available from: https://doi.org/10.1089/end.2021.0542. Epub ahead of print.
23. Young HH. The early diagnosis and radical cure of carcinoma of the prostate. Being a study of 40 cases and presentation of a radical operation which was carried out in four cases. *J Urol*. 2002;168:914–921.
24. Laydner H, Akca O, Autorino R, et al. *J Endourol*. 2014;28(12):1479–1486.
25. Kaouk JH, Akca O, Zargar H, et al. *Urology*. 2016;94–129-38.
26. Garisto J, Bertolo R, Wilson CA, Kaouk J. The evolution and resurgence of perineal prostatectomy in the robotic surgical era. *World J Urol*. 2020;38(4):821–828. Available from: https://doi.org/10.1007/s00345-019-03004-1. Epub 2019 Dec 6.
27. Wiltz AL, Shikanov S, Eggener SE. Robotic radical prostatectomy in overweight and obese patients: oncological and validated-functional outcomes. *Urology*. 2009;73:316–322.

CHAPTER

The evolution of single-port robot-assisted transperineal radical prostatectomy

5

Riccardo Bertolo[1], Juan Garisto[2,3] and Jihad Kaouk[4]

[1]*Department of Urology, San Carlo di Nancy Hospital, Rome, Italy*

[2]*Harvard Medical School, Boston, MA, United States*

[3]*Division of Urology, Veterans Affairs Medical Center, Boston, MA, United States*

[4]*Department of Urology, Glickman Urological and Kidney Institute, Cleveland Clinic, Cleveland, OH, United States*

5.1 Background

Open perineal radical prostatectomy was first described by Young[1] in 1905 and remained the most common approach for surgical treatment of prostate cancer until the mid-1970s. The retropubic technique, initially described by Millin[2] in 1947, was popularized in the early 1980s, thanks to the development of the Walsh anatomic approach.[3] Still during the 1980s, Weiss and colleagues[4] applied the techniques of cavernous nerve sparing described by Walsh and coworkers[3] to open perineal radical prostatectomy. Nowadays, among the surgeons still opting for open surgery in the treatment of localized prostate cancer, the retropubic approach represents the preferred technique by most, although this preference seems to be based more on surgeon preferences rather than on the available evidence.

Robot-assisted radical prostatectomy was first described in Europe in the early 2000s.[5,6] Its use rapidly expanded over the last two decades. Currently, robot-assisted radical prostatectomy performed by transperitoneal approach represents the most performed surgical approach for prostate cancer in the United States.[7]

Laydner et al. first investigated the applicability of the da Vinci Si robotic system (Intuitive Surgical, Sunnyvale, California, United States) for the perineal approach to robot-assisted radical prostatectomy.[8] The authors evaluated the feasibility of perineal robotic radical prostatectomy in preclinical models, then Kaouk leading the same group reported perineal robotic radical prostatectomy successfully completed in four patients who did not require concomitant pelvic lymph-nodes dissection, applying a single port robotic perineal approach by using the da Vinci Si robotic platform.[9]

Again, the Cleveland Clinic group investigated the feasibility of perineal robot-assisted radical prostatectomy on a cadaver model.[8] The novel SP1098 da Vinci robotic platform was used.[10] The use of the single-port purpose-built

Single-Port Robotic Surgery in Urology. DOI: https://doi.org/10.1016/B978-0-323-91906-7.00012-0

robotic platform allowed the authors to describe, for the first time, the feasibility of robotic perineal pelvic lymph-nodes dissection.[11] In parallel, Tuğcu and colleagues described "The Tugcu Bakirkoy" perineal robotic radical prostatectomy with pelvic lymph-nodes dissection in seven patients, using the da Vinci Xi HD Surgical System (Intuitive Surgical, Inc., Sunnyvale, California, United States) on single Gel-port platform.[12]

5.2 Surgical technique for perineal robotic radical prostatectomy with single-site multiport approach (R-LESS)

5.2.1 Instruments armamentarium

To achieve the perineal access, scalpel, tissue forceps, Metzenbaum scissors, Allis and right-angle forceps, as well as Richardson retractors are used. A GelPOINT Mini Advanced Access Platform (Applied Medical, Rancho Santa Margarita, California, United States) is used, including an Alexis wound retractor and a GelSeal Cap (Applied Medical) (Fig. 5.1).

The da Vinci Si system is used in a 3-arms configuration. One 12-mm port for the robotic optic, one 10-mm port for the assistant, and two 8-mm robotic ports are inserted through the GelSeal Cap in a diamond-shape configuration. An additional 5-mm port for the assistant can also be placed (Fig. 5.2).

A 0-degree robotic scope is used. Robotic monopolar curved scissors (Hot Shears) and robotic needle driver are used in the right hand. A robotic grasper (ProGrasp forceps) is used in the left hand.

5.2.2 Prostate assessment and patient positioning

The patient is placed in the lithotomy position (Fig. 5.3).

Digital rectal examination, cystoscopy, and transrectal ultrasonography should be preoperatively performed to get an estimate of the prostate size. The patient is taped to the surgical table. The Trendelenburg position can be applied if needed.

5.2.3 Perineal access and robot docking

A 4–5 cm perineal incision is made (Fig. 5.4), the subcutaneous tissue is dissected, and the central tendon is divided (Fig. 5.5).

A procedure analog to the Belt approach is performed, retracting the external sphincter muscle superiorly.[13] The single port device is inserted into the wound, and CO_2 insufflation is obtained. The robot is brought into the field and docked from behind the head (Fig. 5.6).

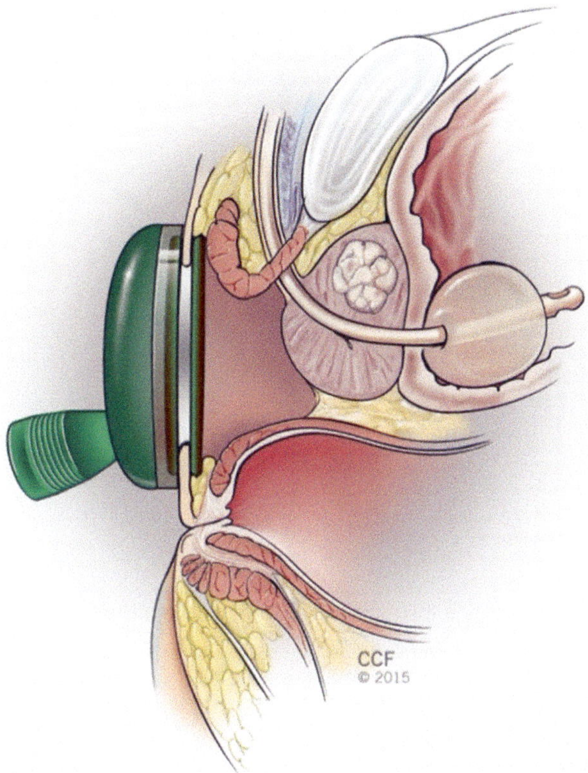

FIGURE 5.1

Schematic drawing illustrating the sagittal view of the placement of single port device after initial dissection.

Enough room for the anesthesiology team to access the head, the neck, and the chest of the patient is available from either side.

5.2.4 Robotic procedure step-by-step

The recto-urethralis muscle is identified and divided. The bedside assistant provides crucial information to the console surgeon while performing digital rectal examination. The levator ani fibers are split aside, and the posterior aspect of the Denonvilliers fascia is identified and incised, preserving the neurovascular bundles intact. The posterior plane of the prostate, the vas deferens, and the seminal vesicles are dissected. The prostatic pedicles are clipped and divided.

The prostatic apex is dissected, the urethra is transected, and the Foley catheter is clipped and cut (Fig. 5.7).

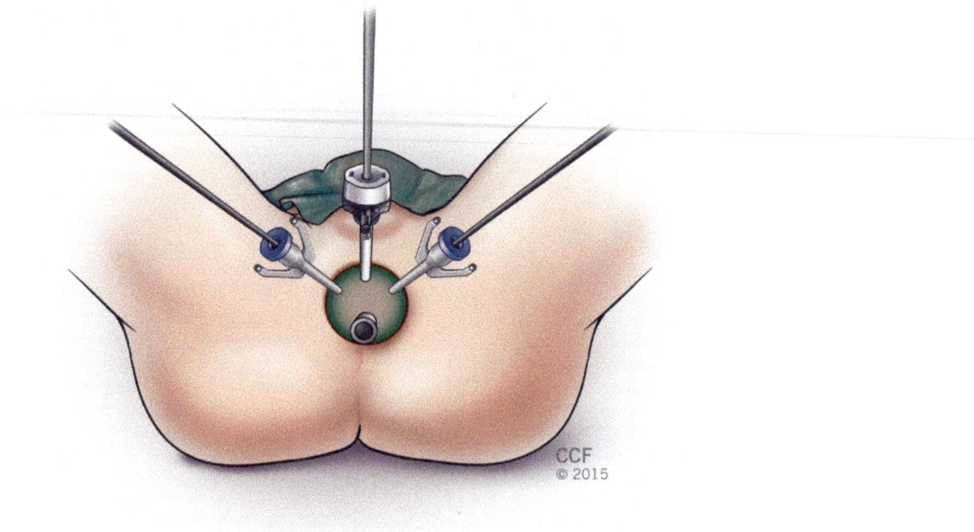

FIGURE 5.2

Schematic drawing illustrating instruments' location through the GelSeal Cap of the GelPOINT Mini Advanced Access Platform.

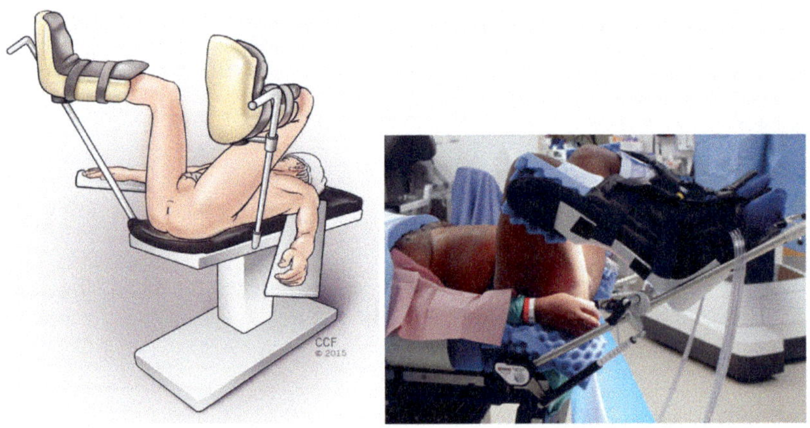

FIGURE 5.3

Patient placed in the lithotomy position for robotic perineal prostatectomy. Note the less than 10 degrees Trendelenburg position.

The proximal portion of the catheter is used as a handle for retraction. The anterior and lateral planes of the prostate are dissected. The bladder neck junction is identified and dissected off from the prostate. The catheter clip is removed to deflate the balloon and the specimen is divided from the bladder (Fig. 5.8).

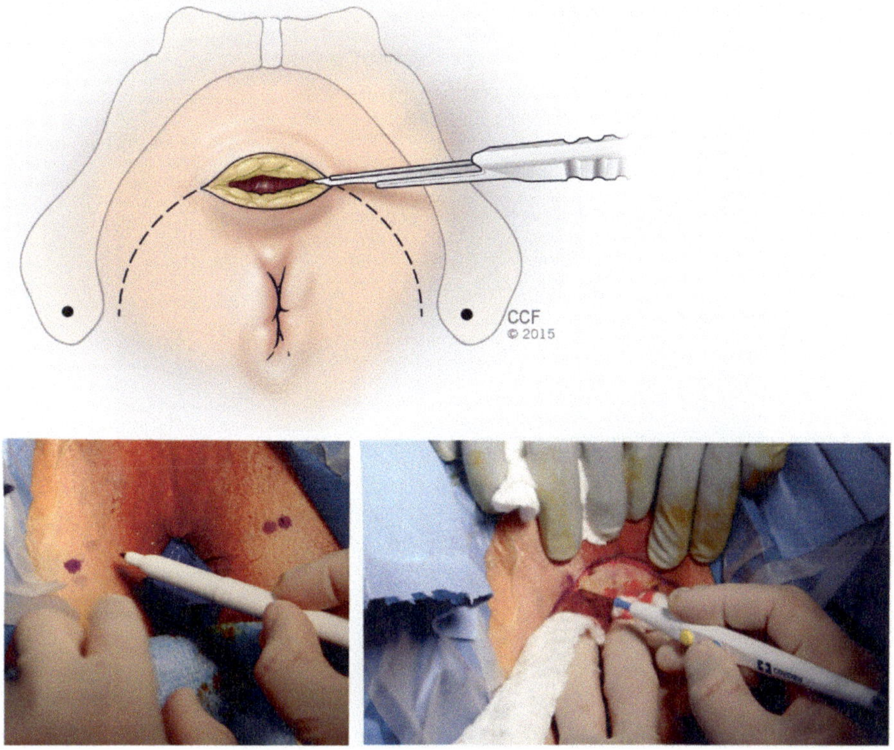

FIGURE 5.4

A 4–5 cm perineal incision is performed at the apex of a semicircular line extending between the ischial tuberosities to get the transperineal access.

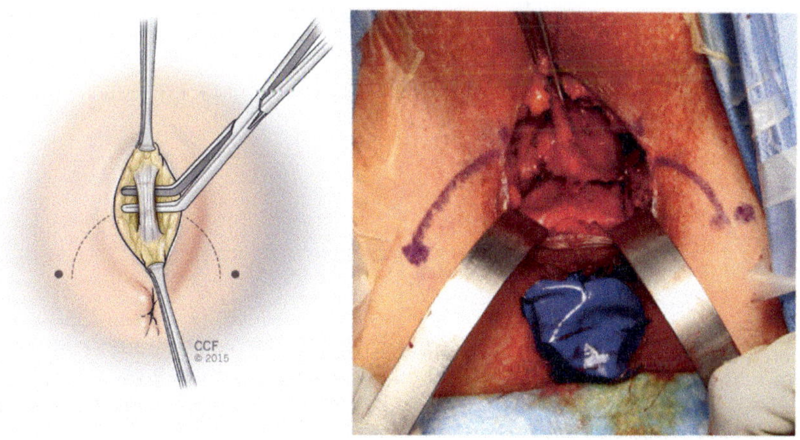

FIGURE 5.5

Division of the central tendon.

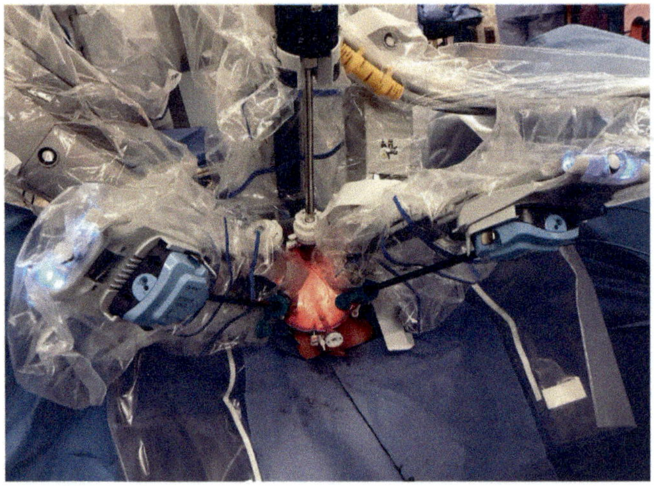

FIGURE 5.6

The multiarm robot docked from behind the head of the patient.

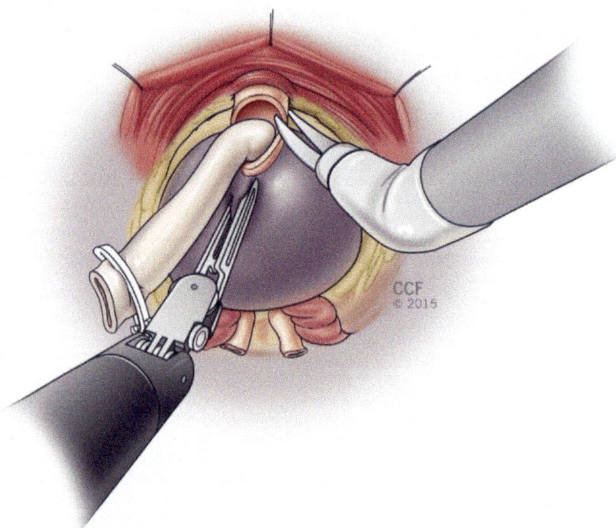

FIGURE 5.7

Transection of the urethra and exposure of the Foley catheter.

A second Foley catheter is inserted to guide the vesicourethral anastomosis, performed according to the Van Velthoven technique[14] (Fig. 5.9).

After completion, the water-tightness of the anastomosis is tested with instillation of 200 mL of saline in the bladder. The robot is undocked, and the

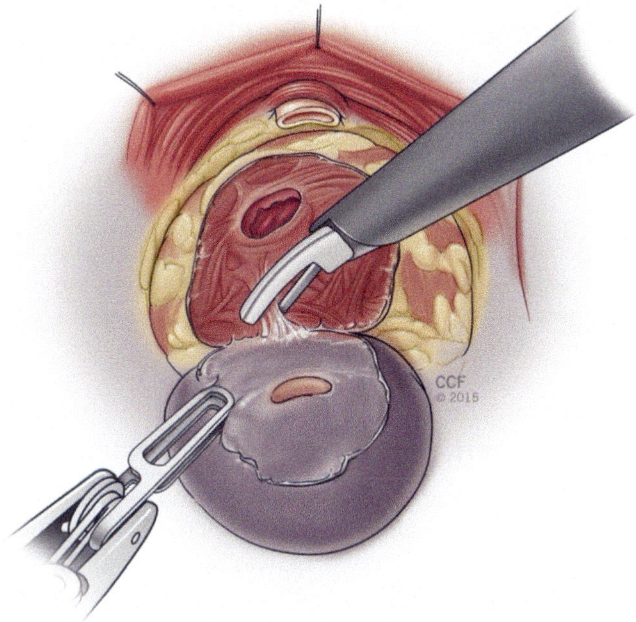

FIGURE 5.8

The prostatectomy is completed: the specimen is separated from the bladder.

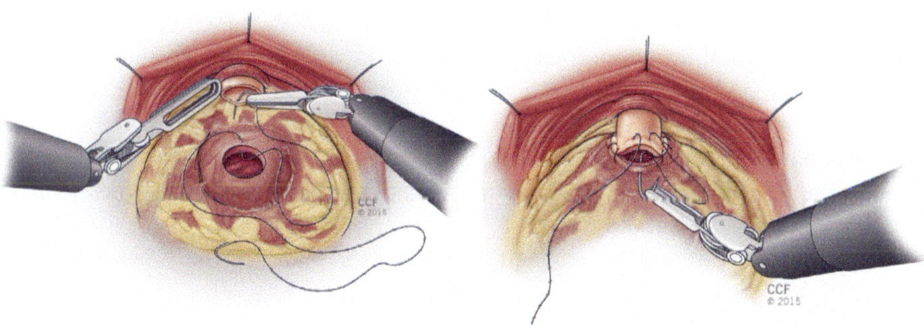

FIGURE 5.9

Vesicourethral anastomosis, performed according to the Van Velthoven technique.

single-port device is removed. The specimen is retrieved through the perineal incision. The perineal fascial and the subcutaneous planes are re-approximated, and the skin incision is sutured.

5.3 Surgical technique for transperineal single-port robot-assisted radical prostatectomy performed by using the single-port surgical platform

The recent development of the purpose-built single-port (SP) robotic platform has prompted the re-discovery of the transperineal approach to radical prostatectomy.

The intracorporeal triangulation of the double-jointed instruments and the fully wristed three-dimensional optics do contribute to resolving the external clashing typical with the previous-generation multiarms robotic platform, allowing surgery to be performed in a narrow surgical field.

5.3.1 Positioning, perineal access, and port placement

After the perineal access is gained as mentioned above for the multiport approach (see Section 5.2) and the GelPOINT device (Applied Medical, Rancho Santa Margarita, CA, United States) is inserted, the 2.5-cm specialized, multichannel single-port trocar is placed. This novel single port accommodates three 6-mm robotic instruments and a 10×12-mm articulating camera (Fig. 5.10).

The robotic device is then docked (Fig. 5.11).

5.3.2 Robotic single-port perineal radical prostatectomy

The technique reproduces the standard steps described for the single-site multiport approach. Once the single-port robotic device is docked, the posterior prostatic space is developed to expose the levator ani muscles on either side of the posterior aspect of the prostate (Fig. 5.12).

The levator ani fibers are split along the lateral aspects of the prostate, and Denonvilliers fascia is opened. This can be performed in a way that preserves the neurovascular bundles bilaterally (Fig. 5.13).

Once Denonvilliers fascia is open, the posterior plane of the prostate is delineated and followed to find the insertion of the seminal vesicles and vas deferens bilaterally (Fig. 5.14).

Once this is done, bilateral vascular pedicles to the prostate are identified and ligated (Fig. 5.15).

The prostatic apex and urethra are dissected, and the urethra is sharply cut to expose the catheter (Fig. 5.16).

The catheter may be clipped to maintain fluid in the catheter balloon and then cut. The proximal portion of the catheter may be used to place traction on the prostate. The junction between the bladder neck and prostate is identified and opened (Fig. 5.17).

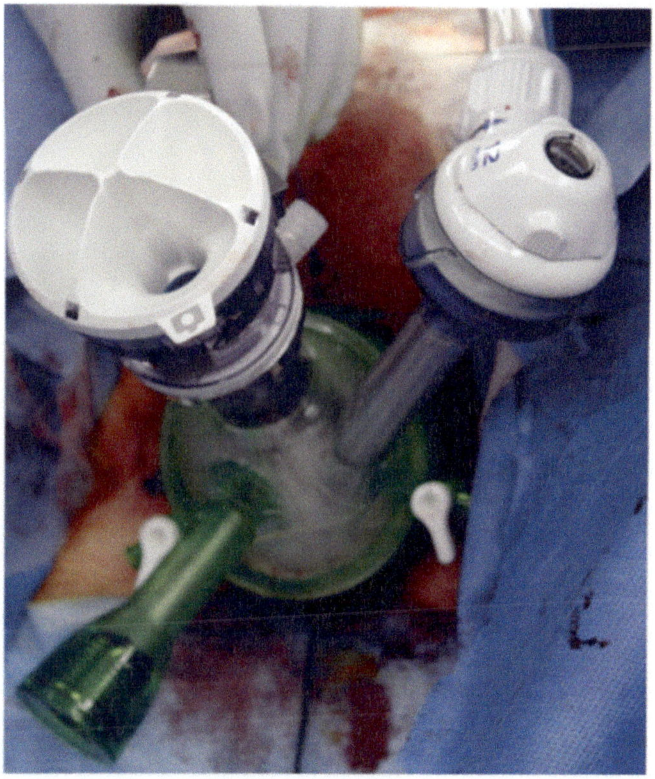

FIGURE 5.10

Details of the 2.5-cm specialized, multichannel, single-port. This novel single port accommodates three 6-mm robotic instruments and a 10 × 12-mm articulating camera. The ports are placed into the GelSeal-Cap prior to assembling it over the external ring of the GelPOINT platform.

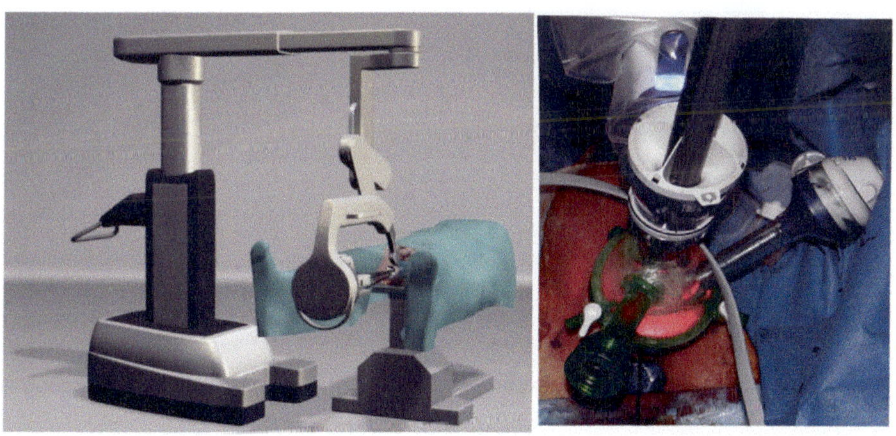

FIGURE 5.11

Cartoon and picture of the docking of the robotic platform during robot-assisted single-port transperineal radical prostatectomy.

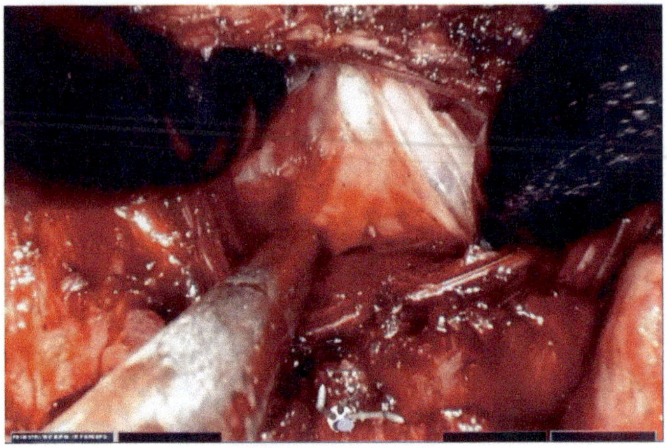

FIGURE 5.12

Development of the peri-rectal space during robot-assisted single-port transperineal radical prostatectomy.

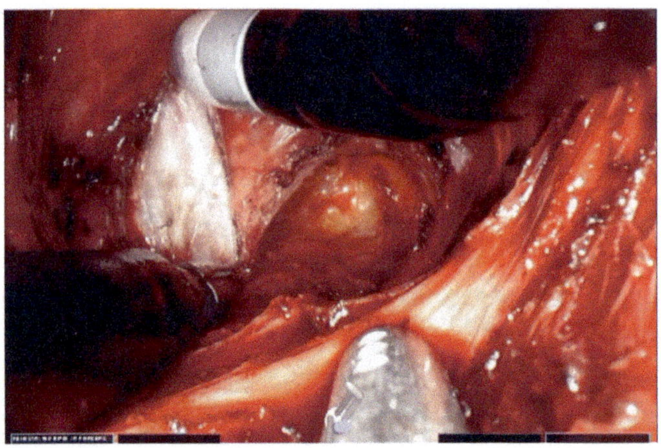

FIGURE 5.13

Dissection of the Denonviller's fascia during robot-assisted single-port transperineal radical prostatectomy (DF: Denonviller's fascia; PF: peri-rectal fat; R: rectum).

Once this is done, the clip on the catheter is removed to allow emptying of the catheter balloon: prostatectomy is completed. The vesicourethral anastomosis is then performed according to the Van Velthoven technique, as summarized in Fig. 5.9 for the multiarms single-port approach (Fig. 5.18).

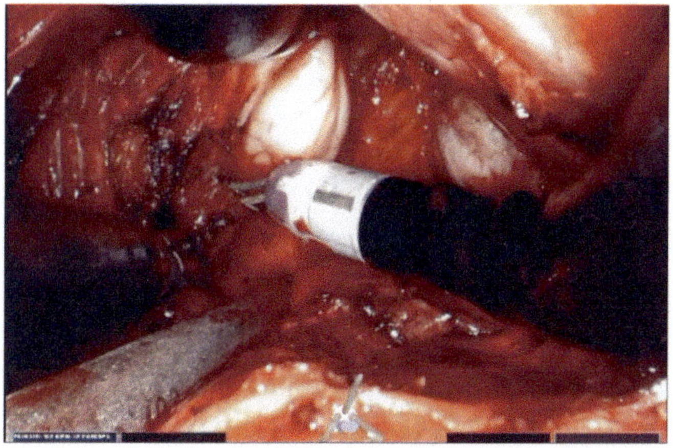

FIGURE 5.14

Dissection of the seminal vesicles during robot-assisted single-port transperineal radical prostatectomy.

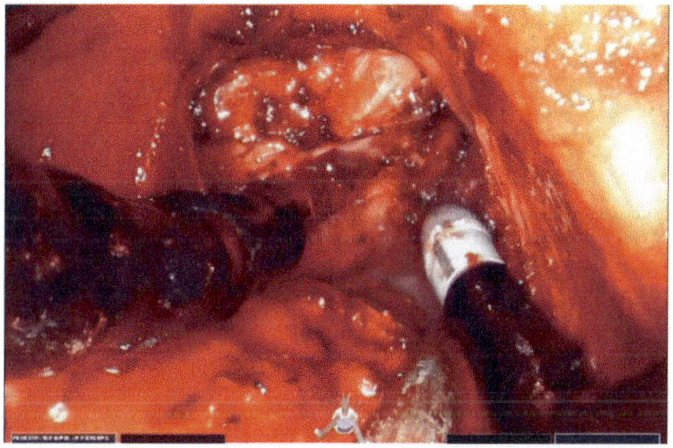

FIGURE 5.15

Dissection of the prostatic lobes during robot-assisted single-port transperineal radical prostatectomy.

5.3.3 Pelvic lymph-nodes dissection

Specifically for the single-port approach with the SP platform, the technique for transperineal pelvic lymph-nodes dissection is briefly described. Once prostatectomy is completed, access to bilateral pelvic lymph nodes is performed by developing the inferior lateral perivesical space that was initially entered after the splitting

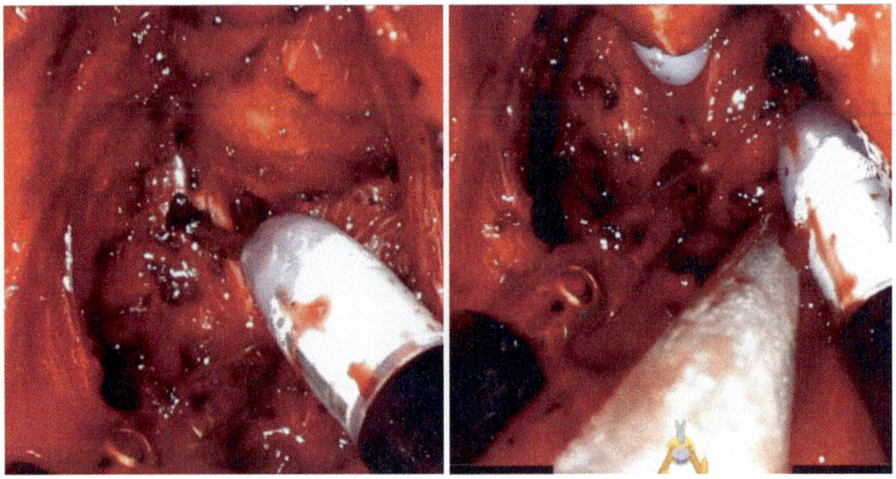

FIGURE 5.16

Apical dissection: division of the urethral stump during robot-assisted single-port transperineal radical prostatectomy.

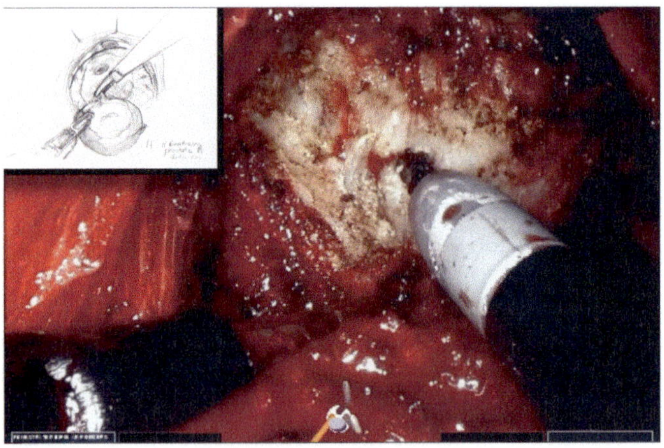

FIGURE 5.17

Bladder neck dissection during robot-assisted single-port transperineal radical prostatectomy.

of the levator ani muscle on either side (see Fig. 5.13). The bladder is retracted medially with the use of the ProGrasp forceps to expose the obturator fossa and the inferior edge of the external iliac vein. From this approach, the obturator fossa and external iliac vessels are visualized. The single-port access port allows easy access to the obturator and external iliac nodal chains on either side, while maintaining the

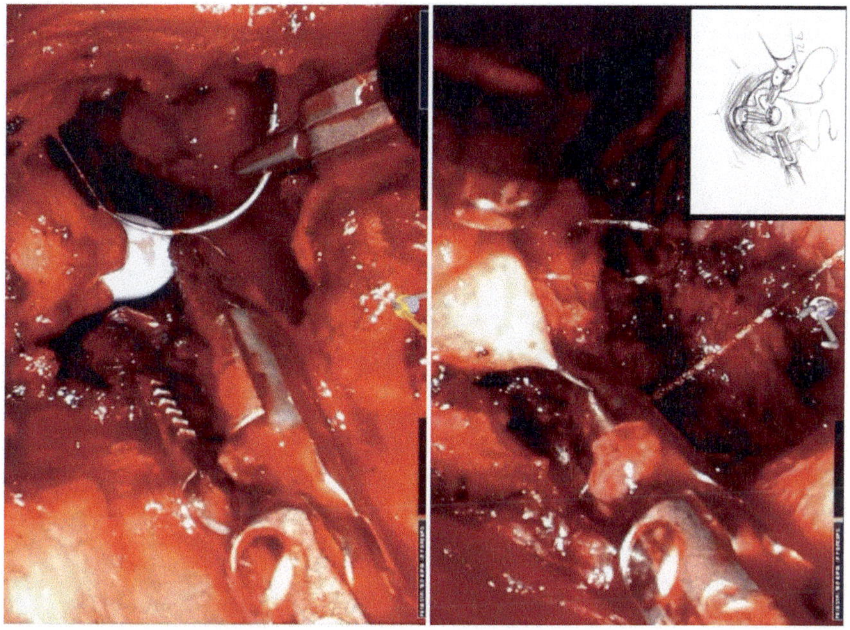

FIGURE 5.18

Urethro-vesical anastomosis performed during robot-assisted single-port transperineal radical prostatectomy.

exposure and the triangulation of the robotic instruments. The lymphatic tissue in the obturator fossa is removed initially to improve visualization of the inferior edge of the external iliac vessels, followed by excision of the external iliac nodal chain. It is important to remember that the anatomy is reversed from the typical retropubic approach, meaning that the obturator structures will be encountered before encountering the external iliac vein (Figs. 5.19 and 5.20).

5.4 Discussion

Single-port surgery has the potential to improve the patient satisfaction, with earlier convalescence, better pain control and improved cosmesis.[15] While there has been increased implementation of laparo-endoscopic single-site surgery at major tertiary centers, sizeable challenges have still existed with this technique, including the clashing of laparoscopic instruments, the adverse ergonomics, and the inability to recreate the intraabdominal triangulation around the surgical field. A new da Vinci single-port robotic system has been designed specifically to overcome these adversities. While the older Si and Xi da Vinci platforms have required multiple trocars

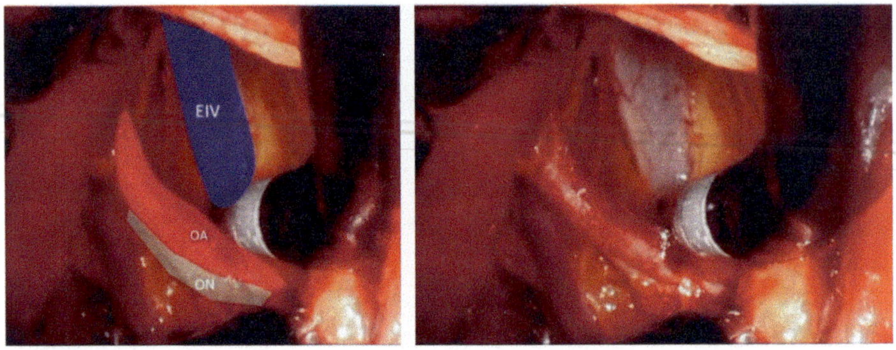

FIGURE 5.19

Right side pelvic lymph-nodes dissection during robot-assisted single-port transperineal radical prostatectomy. The reader should note that the anatomy is reversed from the typical retropubic approach: the obturator structures (ON: obturator nerve; OA: obturator artery) are encountered before encountering the external iliac vein (EIV).

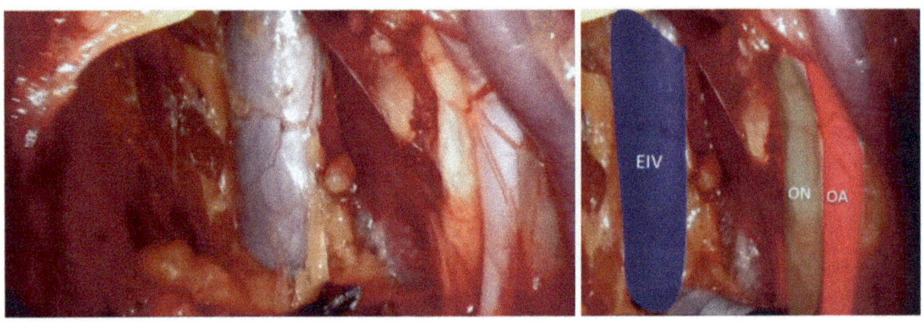

FIGURE 5.20

Left side pelvic lymph-nodes dissection during robot-assisted single-port transperineal radical prostatectomy (ON: obturator nerve; OA: obturator artery; EIV: external iliac vein).

to allow for the introduction of the camera and the devices endoscopically, the novel SP da Vinci platform allows the placement of three EndoWristed robotic instruments and an articulating three-dimensional camera via a specialized 2.5-cm multichannel port. The preservation of the EndoWrist technology, along with introduction of a unique "elbow" joint, allow for the preservation of wristed movements while recreating the intraabdominal triangulation. This platform is actually pushing the envelope of what urologists (and surgeons from other specialties) can achieve via intraabdominal, extraperitoneal and perineal approaches[16] (see Chapter 4).

Specifically regarding the transperineal approach to prostate surgery, the reported advantages have included shorter operative time and hospital stay,[17] lower cost for patients who do not need bilateral pelvic lymph-nodes dissection,[18] lower postoperative incidence of anastomotic stricture and inguinal hernia,[19,20] as well as lower blood loss and transfusion rates.[17,21]

In the perineal approach, there is no need for the penile dorsal vein complex sectioning. Disadvantages were reported to be higher rates of rectal injury[21] and difficult nerve-sparing for large prostates.[22] During the posterior dissection, the bedside assistant has a critical role while performing digital rectal examination and providing feedback to the console surgeon about the correct plane of dissection. This helps to minimize the risk of rectal injury. It is important also, however, to avoid dissecting too anteriorly to avoid inadvertent injury to the bulbar urethra before the identification of the prostate.

Until recently, one of the major disadvantages attributed to perineal radical prostatectomy was that patients requiring bilateral pelvic lymph-nodes dissection needed to have it performed through a separate access, usually via laparoscopy. Saito and Murakami,[23] however, described a technique for pelvic lymph-nodes dissection through the same perineal incision, using several retractors for direct view or laparoscopic assistance. They obtained a median of eight lymph nodes via the perineum in 20 patients who underwent perineal radical prostatectomy.

Keller and colleagues[24] reported extended pelvic lymph-nodes dissection via the same perineal incision in 90 patients who underwent perineal radical prostatectomy. After the prostate removal, they performed the extended pelvic lymph-nodes dissection under direct vision using a self-retaining system, retracting the bladder medially. Their technique yielded the retrieval of 19 lymph nodes on average, with a mean operative time of 150 minutes. Lymphocele developed in seven (7.8%) patients, with four (3.3%) of them needing treatment.

In 2016 Cleveland Clinic group described the technique of robotic pelvic lymph-nodes dissection performed through the same perineal incision after prostatectomy by using the purpose-built robotic platform in the preclinical model. As such, the technology of the SP1098 allowed for the feasibility of even extended lymph-nodes dissection.[11]

Some authors found that robot-assisted radical prostatectomy has lower estimated blood loss and transfusion rates in comparison with open retropubic radical prostatectomy, although the absence of randomized studies still limits definitive conclusions.[25–28] Whether similar advantages could also be observed using the robotic platform specifically for robotic perineal radical prostatectomy is not clear yet and must be further studied. Horuz et al.[29] studied different ultrasonography parameters in 50 patients undergoing open perineal radical prostatectomy to identify those associated with difficulty of the operation, determined by a score composed of operative time, estimated blood loss, and surgeon subjective judgment. They found that a prostate volume > 40 mL and a probe-divergence angle < 45 degrees (the angle between the vertical axis and the oblique plane obtained when directing the probe toward the base of the prostate during suprapubic ultrasonography) were independent predictors for a more difficult procedure. In an endoscope-assisted perineal radical prostatectomy technique, investigators from the same institution described the use of a Collins knife to cut under direct vision the prostate-vesical junction, which is regarded as one of the most difficult steps in radical prostatectomy.[30] In the preclinical experience by Kaouk and coworkers,[11] the only positive margin for prostatic tissue was located exactly at the prostate-vesical junction in one cadaver. This was also the largest prostate (28 cc), while the two other glands were 22 and 25 cc on preoperative transrectal ultrasonography. Because of the small sample size, however, the authors were unable to claim any correlation between prostate size and margin status. Potential technical benefits of robotic transperineal approach to radical prostatectomy over the open one would include the magnified view as well as the long, stable, and articulated robotic instruments, which may help to minimize the difficulty of performing a procedure in such a narrow and sometimes deep surgical field.

It is very recently that Kaouk and coworkers published the first series of transperineal SP robotic prostatectomies, described by Lenfant et al. in 2021.[31] The report included 26 patients with complex surgical histories and high-risk localized prostate cancer with limited alternative therapeutic options. The patients were prospectively entered into an institutional review board—approved database. The data of a recent historical cohort of 86 consecutive patients treated with standard transperitoneal multiports RARP performed by the same surgeon before the beginning of the SP experience, were used as comparators. A matched-paired comparison was made based on age at surgery, prostate-specific antigen level, preoperative Gleason score, and history of abdominal surgery. After matching, baseline characteristics were comparable except for the rate of prior laparotomy, which was higher in the SP group (52% vs 8%, $P < 0.001$). In the SP group, 84% of the patients had a high risk and an unfavorable intermediate risk of positive surgical margins versus 57% in the standard RARP group ($P = 0.03$). While the rate of >3 mm positive margins was higher in the SP group (38.5% vs 7.7%, $P < 0.01$), the number of patients with biochemical recurrence at 1 year was comparable between the approaches. Overall, the study suggested that, compared to the standard, in experienced hands functional and oncological equivalent outcomes are possible with the SP approach at 12 months.

Undoubtedly, to justify the robotic perineal approach, one has to foresee the potential advantages over the standard transabdominal prostatectomy technique. From a purely anatomic standpoint, a perineal approach makes complete sense, because it provides a more direct access to the prostate than the anterior abdominal wall. Potential clinical advantages include the elimination of the first three steps normally performed during the transabdominal robotic approach (bladder mobilization, endopelvic fascia opening, and dorsal vein complex control), which could theoretically result in shorter operative time and lower blood loss.

5.5 Future perspectives

As a completely extraperitoneal approach, transperineal robotic radical prostatectomy virtually eliminates risks of injury to the small bowel or major vessels during the ports-placement, which, despite rarely, can be catastrophic. It also allows avoidance of having to deal with extensive adhesions in patients with previous multiple surgeries or repair of abdominal wall defects with mesh. Obese patients often impose a technical challenge during standard robotic prostatectomy. As such, a greater incidence of case abortion because of pneumoperitoneal pressure and excessive airway pressures has been reported in this group of patients.[32] Although perineal robotic prostatectomy uses CO_2 insufflation to improve visualization, it eliminates the need for pneumoperitoneum and its potential complications, which may make it an attractive alternative for specific patients. The transperineal robotic approach has been described among the available strategies to deal with patients after aborted prostatectomy.[33] Particularly, the value of the transperineal robotic prostatectomy has been reported both for the surgical and the anesthesiology issues when transperitoneal access is not feasible, as well as intraoperative strategies that can be used when there has been prior pelvic instrumentation.

The unique aspects of the novel purpose-built single-port platform could lead to several advantages with respect to the multiarms robotic system. The initial investigation showed promise in furthering the field of single-site/single-port surgery by offering the preservation of triangulation, the EndoWrist technology and the strength of traction via a single keyhole incision. The feasibility of robotic perineal prostatectomy by using the novel purpose-built robotic platform has been demonstrated, including the transperineal–perineal pelvic lymph-nodes dissection.

Future trials are warranted to better define the forthcoming role of this kind of surgery within the field of minimally-invasive urological surgery.

References

1. Young HH. The early diagnosis and radical cure of carcinoma of the prostate: being a study of 40 cases and presentation of radical operation which was carried out in 4 cases. *Bull Johns Hopkins Hosp*. 1905;16:315–321.
2. Millin T. *Retropubic urinary surgery. Carcinoma of the Prostate: Radical Retropubic Prostatectomy*. Baltimore, MD: Williams & Wilkins; 1947;15–17.
3. Walsh PC, Lepor H, Eggleston JC. Radical prostatectomy with preservation of sexual function: anatomical and pathological considerations. *Prostate*. 1983;4:473–485.
4. Weiss JP, Schlecker BA, Wein AJ, Hanno PM. Preservation of periprostatic autonomic nerves during total perineal prostatectomy by intrafascial dissection. *Urology*. 1985;26:160–163.
5. Abbou CC, Hoznek A, Salomon L, et al. Remote laparoscopic radical prostatectomy carried out with a robot. Report of a case. *Prog Urol (Fre)*. 2000;10:520–523.
6. Binder J, Kramer W. Robotically-assisted laparoscopic radical prostatectomy. *BJU Int*. 2001,87:408–410.
7. Lowrance WT, Eastham JA, Savage C, et al. Contemporary open and robotic radical prostatectomy practice patterns among urologists in the United States. *J Urol*. 2012;187:2087–2092.

8. Laydner H, Akça O, Autorino R, et al. Perineal robot-assisted laparoscopic radical prostatectomy: feasibility study in the cadaver model. *J Endourol*. 2014;28(12):1479–1486. Available from: https://doi.org/10.1089/end.2014.0244.
9. Kaouk JH, Akca O, Zargar H, et al. Descriptive technique and initial results for robotic radical perineal prostatectomy. *Urology*. 2016;94:129–138. Available from: https://doi.org/10.1016/j.urology.2016.02.063. Epub 2016 May 24.
10. Bertolo R, Garisto J, Gettman M, Kaouk J. Novel system for robotic single-port surgery: feasibility and state of the art in urology. *Eur Urol Focus*. 2018;pii:S2405–4569(18):30152-4. Available from: https://doi.org/10.1016/j.euf.2018.06.004 [Epub ahead of print].
11. Ramirez D, Maurice MJ, Kaouk JH. Robotic perineal radical prostatectomy and pelvic lymph node dissection using a purpose-built single-port robotic platform. *BJU Int*. 2016;118(5):829–833. Available from: https://doi.org/10.1111/bju.13581. Epub 2016 Sep 3.
12. Tuğcu V, Akça O, Şimşek A, et al. Robotic perineal radical prostatectomy and robotic pelvic lymph node dissection via a perineal approach: the Tugcu Bakirkoy technique. *Turk J Urol*. 2018;44(2):114–118. Available from: https://doi.org/10.5152/tud.2018.24603. Epub 2018 Mar 1.
13. Belt E. Radical perineal prostatectomy in early carcinoma of the prostate. *J Urol*. 1942;78:287.
14. Van Velthoven RF, Ahlering TE, Peltier A, Skarecky DW, Clayman RV. Technique for laparoscopic running urethrovesical anastomosis:the single knot method. *Urology*. 2003;61(4):699–702.
15. Kaouk J, Garisto J, Eltemamy M, Bertolo R. Single-port robotic intracorporeal ileal conduit urinary diversion during radical cystectomy using the SP surgical system: step-by-step technique. *Urology*. 2019;130:196–200. Available from: https://doi.org/10.1016/j.urology.2019.03.023. Epub 2019 Apr 6.
16. Kaouk J, Garisto J, Bertolo R. Different approaches to the prostate: the upcoming role of a purpose-built single-port robotic system. *Arab J Urol*. 2018;16(3):302–306. Available from: https://doi.org/10.1016/j.aju.2018.04.003.
17. Sullivan LD, Weir MJ, Kinahan JF, Taylor DL. A comparison of the relative merits of radical perineal and radical retropubic prostatectomy. *BJU Int*. 2000;85:95–100.
18. Silverstein AD, Weizer AZ, Dowell JM, et al. Cost com- parison of radical retropubic and radical perineal prosta- tectomy: single institution experience. *Urology*. 2004;63:746–750.
19. Gillitzer R, Thomas C, Wiesner C, et al. Single center comparison of anastomotic strictures after radical perineal and radical retropubic prostatectomy. *Urology*. 2010;76:417–422.
20. Matsubara A, Yoneda T, Nakamoto T, et al. Inguinal hernia after radical perineal prostatectomy: comparison with the retropubic approach. *Urology*. 2007;70:1152–1156.
21. Lance RS, Freidrichs PA, Kane C, et al. A comparison of radical retropubic with perineal prostatectomy for localized prostate cancer within the uniformed services urology research group. *BJU Int*. 2001;87:61–65.
22. Janoff DM, Parra RO. Contemporary appraisal of radical perineal prostatectomy. *J Urol*. 2005;173:1863–1870.
23. Saito S, Murakami G. Radical perineal prostatectomy: a novel approach for lymphadenectomy from perineal inci- sion. *J Urol*. 2003;170:1298–1300.
24. Keller H, Lehmann J, Beier J. Radical perineal prosta- tectomy and simultaneous extended pelvic lymph node dissection via the same incision. *Eur Urol*. 2007;52:384–388.
25. Menon M, Tewari A, Baize B, et al. Prospective comparison of radical retropubic prostatectomy and robot-assisted anatomic prostatectomy: the Vattikuti Urology Institute experience. *Urology*. 2002;60:864–868.

26. Coelho RF, Rocco B, Patel MB, et al. Retropubic, laparoscopic, and robot-assisted radical prostatectomy: a critical review of outcomes reported by high-volume centers. *J Endourol*. 2010;24:2003–2015.
27. Kordan Y, Barocas DA, Altamar HO, et al. Comparison of transfusion requirements between open and robotic-assisted laparoscopic radical prostatectomy. *BJU Int*. 2010;106:1036–1040.
28. Tewari A, Sooriakumaran P, Bloch DA, et al. Positive surgical margin and perioperative complication rates of primary surgical treatments for prostate cancer: a systematic review and *meta*-analysis comparing retropubic, laparoscopic, and robotic prostatectomy. *Eur Urol*. 2012;62:1–15.
29. Horuz R, Goktaxs C, Cxetinel CA, et al. Simple preoperative parameters to assess technical difficulty during a radical perineal prostatectomy. *Int Urol Nephrol*. 2013;45:129–133.
30. Albayrak S, Canguven O, Aydemir H, et al. Endoscope- assisted radical perineal prostatectomy. *J Endourol*. 2010;24:527–530.
31. Lenfant L, Garisto J, Sawczyn G, et al. Robot-assisted radical prostatectomy using single-port perineal approach: technique and single-surgeon matched-paired comparative outcomes. *Eur Urol*. 2021;79(3):384–392. Available from: https://doi.org/10.1016/j.eururo.2020.12.013. Epub 2020 Dec 21.
32. Wiltz AL, Shikanov S, Eggener SE, et al. Robotic radical prostatectomy in overweight and obese patients: oncological and validated-functional outcomes. *Urology*. 2009;73:316–322.
33. Chavali JS, Garisto J, Bertolo R, Agudelo J, Dagenais J, Kaouk J. Robotic radical prostatectomy after aborted prostatectomy: still feasible? The experience from a tertiary care center. *J Robot Surg*. 2018. Available from: https://doi.org/10.1007/s11701-018-0870-x [Epub ahead of print].

CHAPTER

Robotic single-port surgery for kidney and upper urinary tract cancers

6

Paul Kogan[1] **and Ronney Abaza**[2]

[1]*Division of Urology, Department of Surgery, Washington University in St. Louis, St. Louis, MO, United States*

[2]*Central Ohio Urology Group, Mount Carmel St. Ann's Hospital, Westerville, OH, United States*

6.1 Background

Gettman et al. described the first cases of single site laparoscopic renal surgery in a porcine model performed via transvaginal approach.[1] This was a natural orifice transluminal endoscopic surgery (NOTES). The first patients who then underwent laparoendoscopic single site (LESS) renal surgery were in 2007 via three adjacent umbilical ports with articulating instruments.[2] As the umbilicus represents an embryonic natural orifice, this was characterized as E-NOTES. These cases consisted of simple and radical nephrectomies for nonfunctioning kidneys and central renal mass not amenable to partial nephrectomy. Internal and external collisions were problematic and, with the exception of several high volume institutions, wide-spread adoption was not seen due to the technical difficulty and high degree of expertise required. The multiport robotic platform was then utilized for the first robotic single-port renal surgery by Kaouk et al. for pyeloplasty and radical nephrectomy via transumbilical approach and a multichannel gel port.[3] This approach, described as robotic-LESS (R-LESS), experienced similar limited utilization to LESS primarily by a few high volume institutions due to the limited mobility, collisions, and challenges of incorporating the fourth robotic arm. In addition, there were reports of lower rates of optimal surgical outcomes. For example, for partial nephrectomy, the trifecta of warm ischemia <20 minutes, negative surgical margins, and the absence of surgical complications was achieved in only 25.6% of patients compared with 42.7% of patients undergoing multiport partial nephrectomy.[4] Nevertheless, the practitioners of LESS surgery reported shorter convalescence, less pain, and improved cosmesis thereby allowing them to persevere through the technical challenges.

The Da Vinci SP was the first purpose-built single-port robotic platform utilizing a multichannel port and articulating endoscope allowing a total of four robotic arms. The feasibility of performing radical and partial nephrectomy with this platform was first described in three cadaver studies.[5] Since its commercial release in 2019,

Single-Port Robotic Surgery in Urology. DOI: https://doi.org/10.1016/B978-0-323-91906-7.00004-1

multiple institutions and surgical groups have adapted this platform for renal surgery including robotic nephrectomy, partial nephrectomy, pyeloplasty, ureteral reconstruction, and nephroureterectomy in both transperitoneal and retroperitoneal approaches.

In our experience with this modality, there is decreased postoperative pain allowing earlier hospital discharge.[6] While there is a distinct learning curve with the SP platform, it is steeper and faster than that of prior single site renal surgeries particularly among those already experienced with robotic renal surgery.

6.2 Positioning

For transperitoneal cases, it is our preference to position patients in a full 90° lateral decubitus position. An axillary roll is placed to protect the brachial plexus. Upper arm is positioned on an arm rest nearly parallel with the lower arm. Kidney rest and bed flexion are not utilized. For retroperitoneal cases, however, bed flexion is utilized to maximize the working space between the lower costal margin and iliac crest. Complete lateral decubitus facilitates gravity-assisted colon mobilization. For nephroureterectomy, it may be beneficial to place the bed in slight Trendelenburg so as to bring the bowel out of the pelvis for the distal ureteral/bladder cuff dissection.

6.3 Single-port site

Using the Hasson technique, a 2.5 fascial opening is made for the port site. The cannula with blunt obturator is placed directly through this. Alternatively, the incision can be extended slightly to accommodate adjacent additional ports and instruments. These can be placed directly through the same incision or through a wound protector with cap such as the GelPORT (Applied Medical, Rancho Santa Margarita, CA) as seen in Fig. 6.1.

Placing the port site at the umbilicus provides a muscle sparing incision that can be extended vertically for specimen extraction as well as partially concealed by the natural contour of the navel. This is generally sufficient for most renal cases. One specific considerations of the SP system is the 10 cm working distance needed from the port site and target anatomy. If this distance is too short then the instruments are unable to fan out and utilize their complete range of motion. This is more common in patients with a small body habitus or with large lower pole tumors approached retroperitoneally. It can be managed with two solutions. One is to move the port site further away from the target anatomy but this is not feasible with large lower pole tumors during the retroperitoneal approach. The other, dubbed "floating dock," is where the seal of the wound protector system, through which the trocars are placed, is moved further from the skin level.

The assistant may be placed adjacent to the primary cannula. However, they will be "handcuffed" to the primary trocar and limited in their movements.

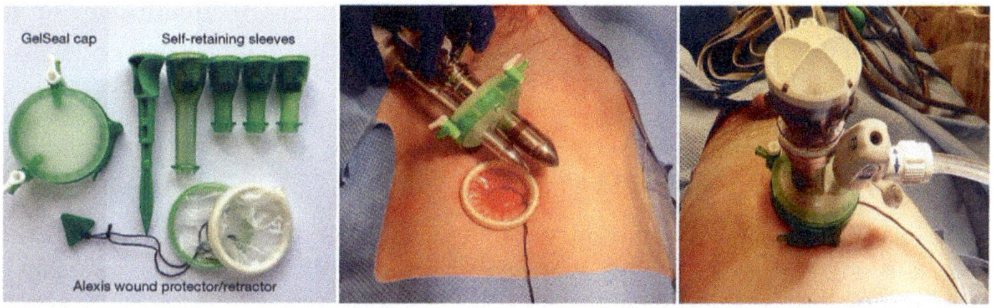

FIGURE 6.1

SP GelPORT configuration with AirSeal cannula. Left: components of the system with bottom right being wound protector, top left cap and top right included port options with obturator. Middle: preparation of the cap with ports performed prior to securing cap as seen on the right.

Left: Appliedmedical.com.

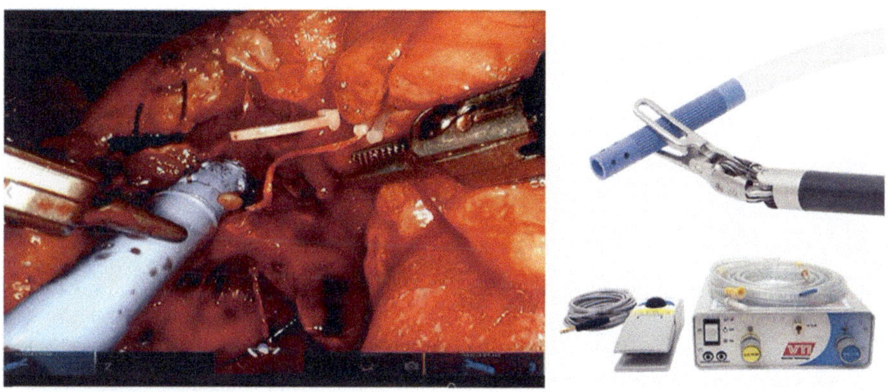

FIGURE 6.2

Left: Intraoperative view of ROSI catheter. This catheter is ideally suited for SP cases where rigid laparoscopic suction requires separate skin incision for optimal use. Top right: example of catheter control with robotic instrument. Bottom right: foot pedal, tubing, and control box for suction system.

Right: VTI-online.com.

Actions such as suture placement and removal may be performed but more complex tasks such as suction and retraction are less feasible through this approach. A " + 1" assistant port can be placed through a separate skin incision; however, this no longer becomes truly single port. This can be avoided with the use of a flexible suction catheter controlled by the surgeon such as the purpose-built Remotely Operated Suction Irrigation System (ROSI, VTI, Nashua, NH), or a nasogastric tube or something similar as seen in Fig. 6.2. This is generally placed through the adjacent assistant port through the same incision or through the fourth port on the main cannula if only three instruments including the camera are needed.

6.4 Insufflation options

CO_2 insufflation can be achieved through several approaches as can be seen in Fig. 6.3. The SP trocar has a luer-lok insufflation through which gas can be instilled. If the GelPoint device is utilized, this too has a similar port. If AirSeal (Conmed, Largo, FL) insufflation is preferred, as is our preference, the trocar can be placed through the GelPoint seal. Trocar size is determined by the size of needles and clamps needed to pass in and out with 8 mm generally being sufficient (Fig. 6.3).

6.5 Instruments

At the present time, the instrument catalog of SP instruments is more limited in its scope than the multiport platform. The most common instruments such as

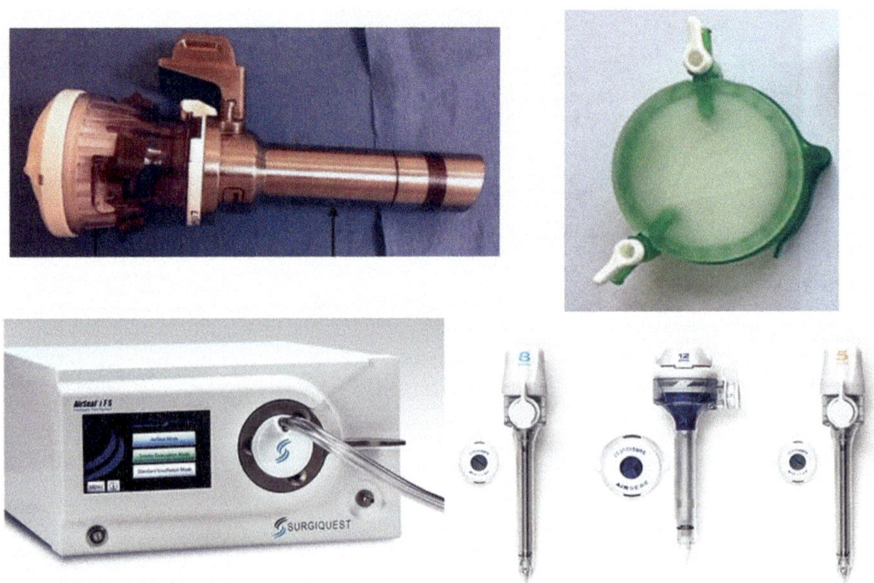

FIGURE 6.3

Insufflation options. CO_2 tubing may be attached directly to the primary cannula (top left), GelSEAL (top right) or through AirSeal trocar (bottom right) placed through GelSEAL.

Top right: Appliedmedical.com; Bottom: Conmed.com.

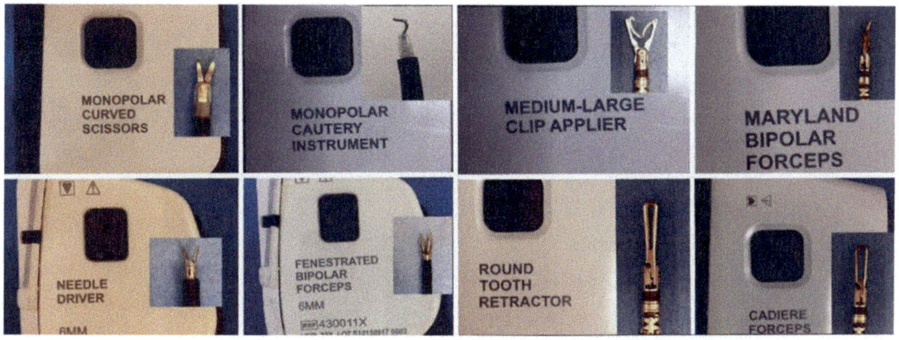

FIGURE 6.4

Several currently available SP instruments. The catalog is more limited than those available for the multiport platforms.

needle drivers, monopolar scissors, and bipolar cautery are available as can be seen in Fig. 6.4. Instruments which are not currently found are the ProGrasp forceps, advanced energy devices such as the Vessel-sealer and large clip applier. In addition, the near-infrared fluorescence (NIRF) is not currently available on the endoscope.

6.6 SP surgery-specific considerations

Below are SP-specific considerations for surgeries compared with multiport approaches. They are not meant to be comprehensive reviews of surgical steps but rather complements for those that are already facile with the procedure.

Radical nephrectomy is most commonly performed via transperitoneal approach unless abdominal adhesions from prior surgery precludes this. The hilum can be controlled with a series of vascular Hem-o-Lok clips. Of note, the clip applier currently available on the SP platform is smaller than the 8 mm used on the multiport. If the main renal artery is felt to be too large for the clip then the segmental renal arteries may be clipped and transected instead. If the hilum is to be controlled with a stapler, which is not this authors preference, then a separate assistant port would be needed for the laparoscopic stapler as a robotic stapler option is not currently available on the SP platform. For extraction, the robot is undocked and specimen is placed into a retrieval bag by the surgeon laparoscopically. Fig. 6.5 demonstrates the cosmetic appearance following extraction where the umbilicus provides excellent concealment. If simple nephrectomy is performed for benign disease then the specimen can be morecellated within the extraction bag to further minimize incision length.

Partial nephrectomy can be performed with the SP platform similar to the multiport platform. As previously mention, the NIRF feature is not available. As

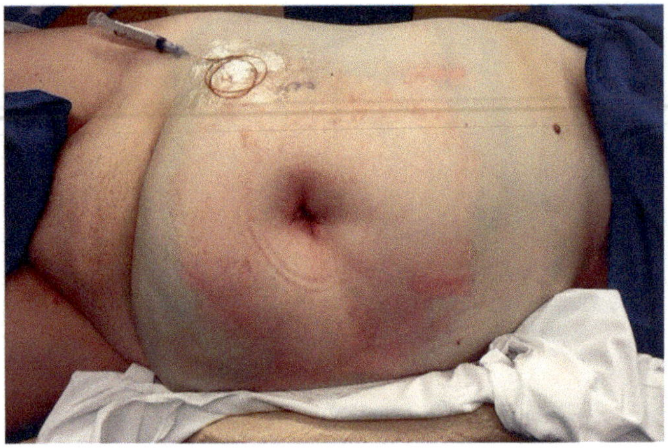

FIGURE 6.5

Photograph of closed incision following extraction of radical nephrectomy specimen using SP platform. Top left catheter routinely used for elastomeric pain pump for local analgesic instillation.

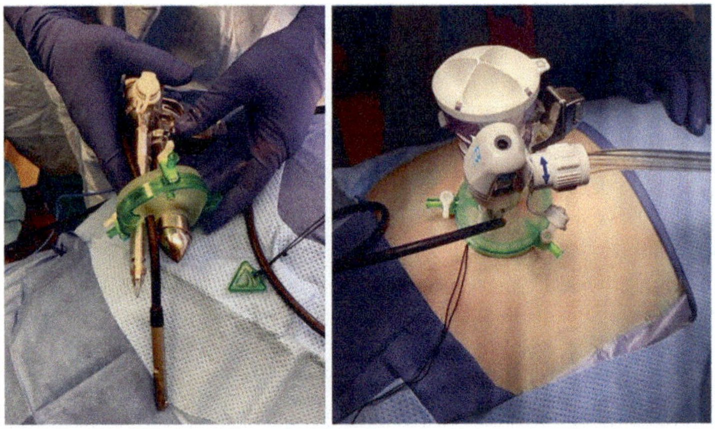

FIGURE 6.6

Port configuration for SP robotic partial nephrectomy. Left: ultrasound probe is preplaced through stab incision in the GelPORT adjacent to AirSeal and robotic ports.

such, alternative measures may be deployed to confirm vascular control and negative margins need to be deployed. We routinely use intraoperative ultrasound to confirm appropriate vascular control and tumor extent. The wound protector easily facilitates additional ports to be placed such as the ultrasound probe as seen in Fig. 6.6, suction catheter, and needle passage in and out of the abdomen. In experienced hands, complex tumors can be resected with short warm ischemia times

(WITs) and excellent outcomes. Fig. 6.7 demonstrates an example of these cases. As mentioned previously, the retroperitoneal approach can be taken as well with identical instruments and utilization of the GelPORT. Fig. 6.8 demonstrates closed extraction site following retroperitoneal partial nephrectomy.

Pyeloplasty is ideally suited for the SP platform given that it does not require extraction of a large specimen and frequently occurs in younger patients where cosmesis is of greater concern. A GelPORT device is not routinely needed and care is taken to not place the incision too close to the target anatomy. We routinely pre-place the stent prior to the robotic portion; however, if this is done after transection

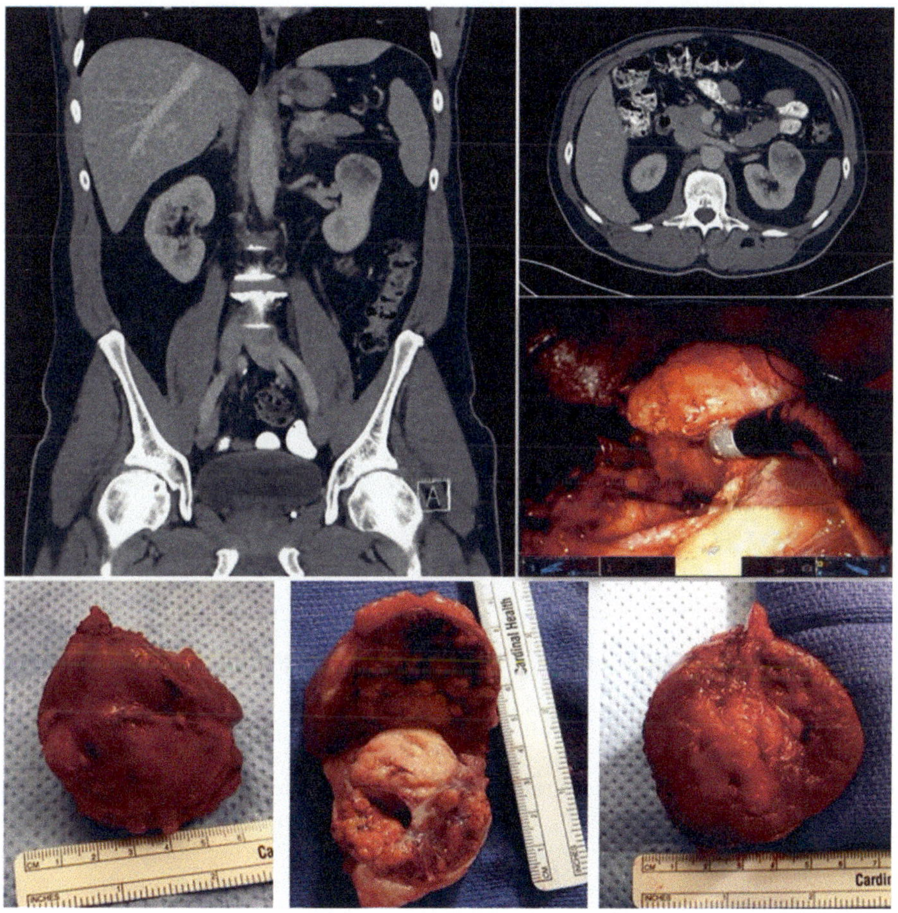

FIGURE 6.7

6 cm left upper pole mass treated with partial nephrectomy using SP platform. WIT was 14:18 min. Top left and top right: preoperative CT scan. Right middle: intraoperative view. Bottom panel from left to right: external, bivalved, and margin views of gross specimen.

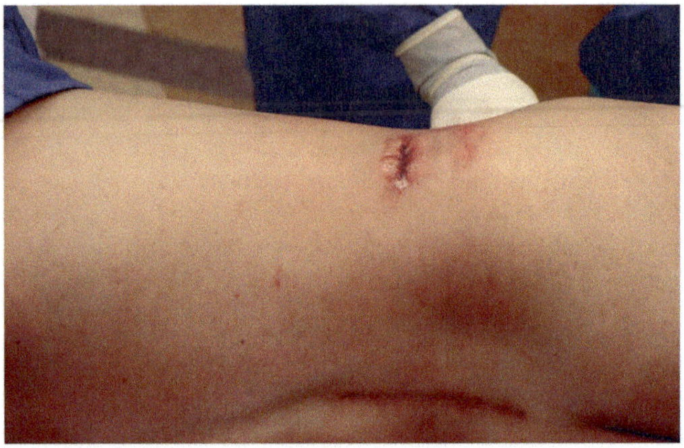

FIGURE 6.8

Postoperative view of closed incision following specimen extraction of a retroperitoneal SP robotic partial nephrectomy.

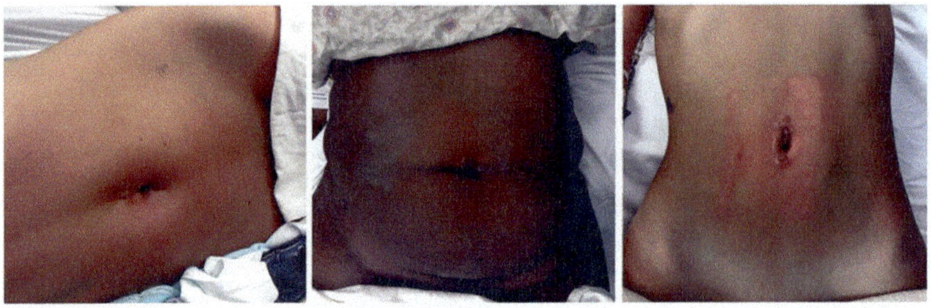

FIGURE 6.9

Skin incisions following closure after SP robotic pyeloplasty. The cannula port alone is used without the GelPORT given the absence of large extraction site or need for advanced instrumentation.

of the ureteropelvic junction, the stent may be passed through one of the port sites within the main cannula or through a large bore angiocatheter elsewhere on the abdomen. Fig. 6.9 illustrates the excellent cosmesis from this approach.

Nephroureterectomy performed with the SP highlights the platforms ability to perform multiquadrant surgery without collision. The multiport platforms can perform this with either redocking the robot, placing additional ports for the different quadrants or with diligent port placement. These steps are not required for the SP, which can simply rotate to the other quadrant from the same position. Fig. 6.10 demonstrates the closed incision following extraction after SP nephroureterectomy. Adjacent to it is catheter for postoperative infiltration of local anesthetic.

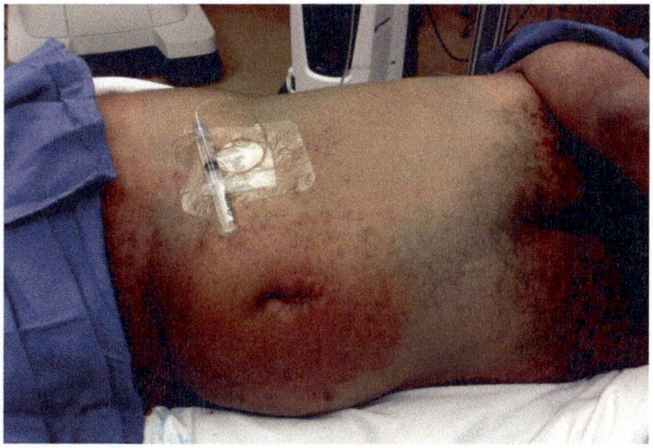

FIGURE 6.10

Skin incision following closure after right SP robotic nephroureterectomy.

6.7 Conclusion

Single site renal surgery has come a long way since its inception. The current SP platform coupled with the ROSI allows true single site renal surgery to be performed on complex cases without compromising on excellent outcomes.

References

1. Gettman MT, Lotan Y, Napper CA, Cadeddu JA. Transvaginal laparoscopic nephrectomy: development and feasibility in the porcine model. *Urology*. 2002;59(3):446–450.
2. Raman JD, Dong S, Levinson A, Samadi D, Scherr DS. Robotic radical prostatectomy: operative technique, outcomes, and learning curve. *JSLS*. 2007;11(1):1–7.
3. Kaouk JH, Goel RK, Haber GP, Crouzet S, Stein RJ. Robotic single-port transumbilical surgery in humans: initial report. *BJU Int*. 2009;103(3):366–369.
4. Komninos C, Shin TY, Tuliao P, Yoon YE, Koo KC, Chang CH, et al. R-LESS partial nephrectomy trifecta outcome is inferior to multiport robotic partial nephrectomy: comparative analysis. *Eur Urol*. 2014;66(3):512–517.
5. Maurice MJ, Ramirez D, Kaouk JH. Robotic laparoendoscopic single-site retroperitioneal renal surgery: initial investigation of a purpose-built single-port surgical system. *Eur Urol*. 2017;71(4):643–647.
6. Abaza R, Murphy C, Bsatee A, Brown Jr. DH, Martinez O. Single-port robotic surgery allows same-day discharge in majority of cases. *Urology*. 2021;148:159–165.

CHAPTER

Robot-assisted single-port radical cystectomy and intracorporeal urinary diversion

7

Fahad Sheckley[1], Michelle Kruse[1], Mubashir Billah[1], Juan Garisto[2,3] and Mutahar Ahmed[1]

[1]*Department of Urology, Hackensack University Medical Center, Hackensack, NJ, United States*

[2]*Division of Urology, Veterans Affairs Medical Center, Boston, MA, United States*

[3]*Harvard Medical School, Boston, MA, United States*

Multiple medical conditions require urinary diversion including radical cystectomy due to bladder cancer, refractory hematuria, radiation cystitis, or chronic pelvic pain. John Simon performed the first urinary diversion in 1851.[1] Sixty years later in 1911, Zaayer described the first ileal conduit. Since then, urinary diversion has become a widely known procedure and despite multiple techniques described in the literature, ileal conduit is the most used technique to this day.[1]

The evolution of robotic techniques has shown significant benefits in terms of postoperative recovery and reduction of complications. The US Food and Drug Administration (FDA) approved the Da Vinci robotic system by Intuitive Surgical Inc. in Sunnyvale, CA in 2000 for use in urologic cases. Robotic surgeries were performed with multiple small incisions in the torso to accommodate the robotic arms. Menon described the first robotic-assisted laparoscopic radical cystectomy in 2003.[2] The robotic approach has dominated the field of radical cystectomy with multiple trials showing no differences in recurrence and overall survival including the RAZOR (Randomized Open versus Robotic Cystectomy) trial comparing the open approach to the robotic-assisted approach.[3,4] While diversion was done externally during this period, recent studies have shown that intracorporeal diversion can be as effective as extracorporeal diversion. Recent papers and metaanalyses have also provided evidence that intracorporeal diversion is more beneficial than external diversion.[5–7] As such, in our institution, we have transitioned to robotic radical cystectomy with intracorporeal diversion. Internal review and published data of our outcomes show similar or improved when compared to current literature.

The robotic approach has evolved overtime with progression of the robotic systems. With the introduction of the Single Port Robot, we have transitioned from multiple small incisions to a single port. The single-port robotic system was

Single-Port Robotic Surgery in Urology. DOI: https://doi.org/10.1016/B978-0-323-91906-7.00002-8

introduced by Intuitive and approved for use by FDA in 2018. Kaouk reported the first use of single-port technology for urinary diversion in 2019.[8] Ahmed provided further description of its use in urinary diversion in 2020.[9,10] Single-port technology is now being used to harvest the bowel for creation of ileal conduits, neobladder, and Indiana pouch whether the native bladder is removed or not.

7.1 Access

A midline incision is made through or above the umbilicus for the robotic access port. The optimal working length of the SP robot instruments is 10–25 cm. There are multiple access platforms that are described for single-port robotic system. In the hrein reported experience by Ahmed et al., the authors initially started with using GelPort (Applied Medical, Rancho Santa Margarita, CA), which is the access platform that was also initially used for multiarms access through a single incision, when using the older generation robotic platforms. It provided good flexibility and prevented pneumoperitoneum. This device enables the concept of a "floating trocar" in which a cap known as "GelSeal" is used to top a sleeve designed to be above the incision housing the trocar of the robotic arms and providing freedom of movement with facilitation of extracorporeal resection and improved articulation of instruments. This allows adjusting the distance of the robot to the target organ. The robot needs to be 10–25 cm from the target organ and floating the trocar allows the ideal positioning and quick intraoperative adjustments.

In more recent days, the authors are using the new SP Access Port Kit (by Intuitive Inc.) (Fig. 7.1), which provides further benefits associated with the

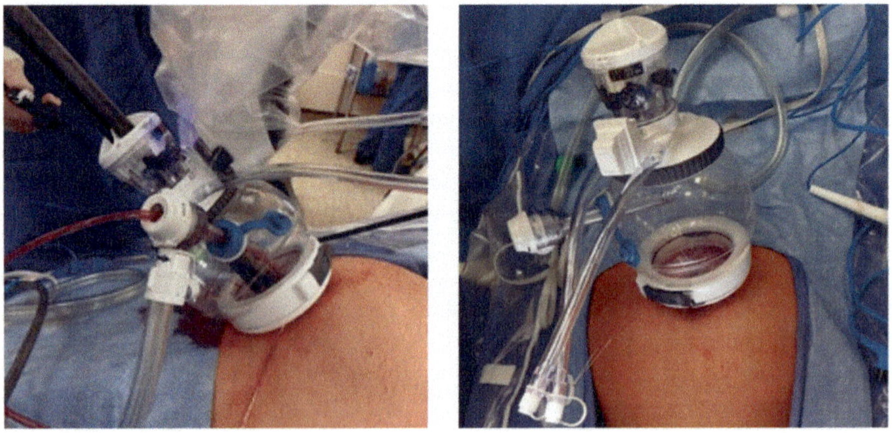

FIGURE 7.1

The use of advanced access techniques in Single-port robotic surgeries. Seen here is the Da Vinci new SP access kit.

single-port system. The distance between trocar and surgical field is usually short but this new access system provides more freedom of movement and reduced incidents of collision among the robotic arms.

This new access system also provides a 12 mm assistant access port and another 8 mm access port on the side of the system. This side port is used to allocate the Airseal (CONMED Corporation, Utica, NY), which is an access management platform that provides stable pneumoperitoneum with smoke evacuation using a valve-free port. It maintains pneumoperitoneum by circulating CO_2 and creating a barrier equal to intraabdominal pressure. Valve-free ports help with providing an access without obstacles. The authors recommend the use of Airseal as it provides stable pneumoperitoneum regardless of rigorous or continuous suction.

Providing a suction technique that is utilized through the single access is also different from the multiport system. Given that there is only one access port, which will be difficult for the bedside assistant to utilize without crowding with the console surgeon control of the arms, a new suction system is introduced. Its name is "Remotely Operative Suction Irrigation System," known as (ROSI), which is composed of a flexible tubing along with a plastic piece at the distal end that can be grasped and moved around using one of the arms of the robot controlled by the console surgeon. If ROSI is not available, flexible suction catheters as is often used by anesthesia for oral airway suction may be used in place of the ROSI system. Other suction techniques have been described like using a urethral catheter with the tip in the pelvis after transecting the urethra controlled by the console surgeon using a robotic arm with suction at the external end of the catheter.[8] An extra assistant port where a regular suction arm by an assistant surgeon can pass is recommended for surgeons with novel experience in the single-port system to provide assistance with retraction and suctioning until the learning curve is achieved.

7.2 Radical cystectomy

The patient is placed in Trendelenburg position. Access is obtained as described earlier. This access allows the robotic trocar to be moved in and out of the body. As a result, the instruments have a wider range of movement, and this range is crucial for reaching deep into the pelvis to perform dissection at the most distal points as well as retracting backwards to retrieve and work with bowel in the proximal abdomen. Instrument positions in the single port are robotic camera at the 12 o'clock position, a monopolar scissor at the 6 o'clock position, a bipolar grasper at the 9 o'clock position, and the Cadiere forceps at the 3 o'clock position. The authors use this instrument configuration but it can be changed as needed. Their preference is to maintain the same position throughout the case to minimize time lost in exchanging instruments.

The procedure begins with incising the peritoneum over the ureter. The ureter is then dissected up proximally and down distally to the level of the bladder sparing the periureteral tissues. The lateral walls of the bladder are then freed following the medial umbilical ligament posteriorly. At this point, the lateral pedicle of the bladder is resected. The ureter is then transected after clipping it at the ureterovesical junction. After the lateral sidewall is dissected and the ureter ligated and the median umbilical ligament are ligated, standard extended lymph nodes dissection proximally from common iliac to distally including the space of Marseille is performed. Posteriorly for male the seminal vesicles and vas is a good landmark to identify and keep them above and avascular plane is mobilize toward the apex of the prostate. The Single-port robot provides additional advantage as the camera angle can be changed according to the need and target relocation helps to move the entire unit of the robot.[11]

In radical cystectomy procedures for men, a combination of bipolar dissector and sharp dissection with scissors is used to remove the prostate and transect the urethra at the space of Retzius. In females, the infundibulopelvic ligament is divided, as the ureter runs underneath it. Females with low volume and no involvement of the posterior bladder, entering in anterior plane of the vagina can preserve the uterus thus configuring the so-called "uterine-sparing" cystectomy. A 12 mm robotic assistant trocar can be placed at the site of the intended stoma at the beginning of the case (Fig. 7.2). For female patients, a 12 mm assistant trocar is optional as the vaginal orifice can be used for bowel stapling.[11]

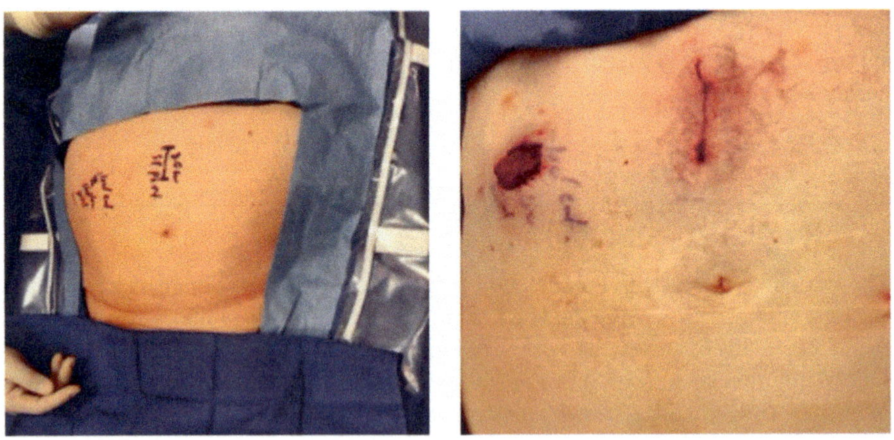

FIGURE 7.2

Site of the single port access midline incision above the umbilicus, marked in the left image and closed in the right image. Also seen is the site of the stoma in the right of the midline incision that can also be used as a site of the assistant port during the surgery prior to maturation of the stoma.

7.3 Harvesting the bowel

Using the length of the robotic instruments, a coarse measurement of the ileum is made, which is harvested about 15–20 cm proximal to the ileocecal junction. For example, 45–50 cm of the ileum is needed to create a neobladder. The distal and proximal ends of the segment are stitched with 3–0 Silk and Vicryl stitches as a mark and as handles for manipulation of the bowel. Staplers are introduced through the vaginal orifice in females (Fig. 7.3), while 12 mm Airseal trocars are commonly used in males (Fig. 7.4).

The ileum is transected on either side, and then the free bowel is anastomosed to the ileum by using 3–0 Vicryl stay sutures to aid in bowel grasping. To create the anastomosis, a second 3–0 Vicryl stitch is placed approximately 6 cm from the free ends of the bowel. Both the jaws of the stapler are advanced through an open end of the bowel to achieve the anastomosis followed by closing the distal end with a stapler in addition to closing it with 3–0 Vicryl sutures.

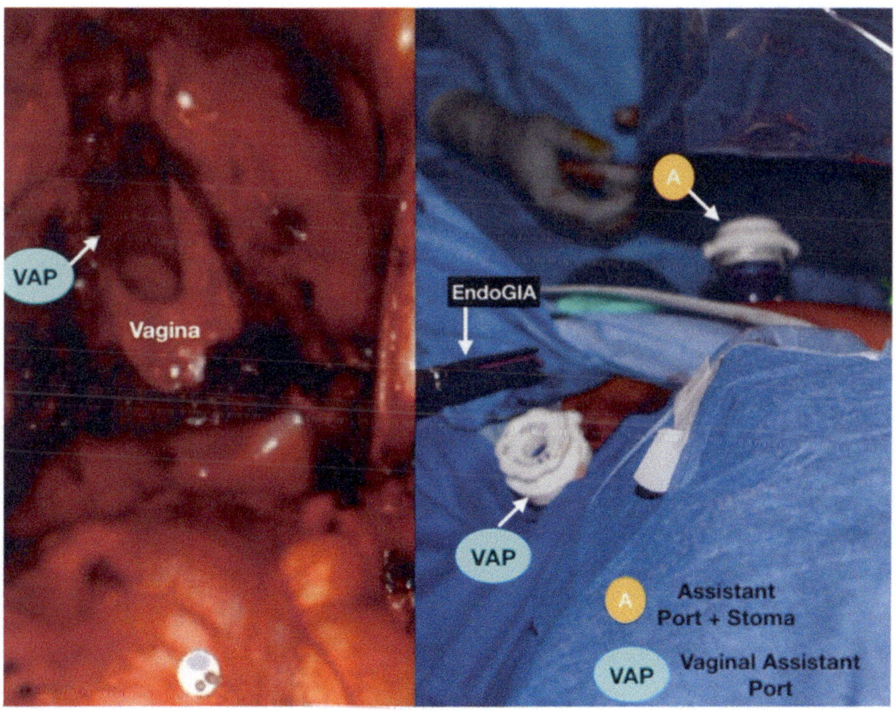

FIGURE 7.3

Intracorporeal bowel harvesting prior to perform the urinary diversion in female. A 12-mm trocar is placed into the vagina for bowel stapling.

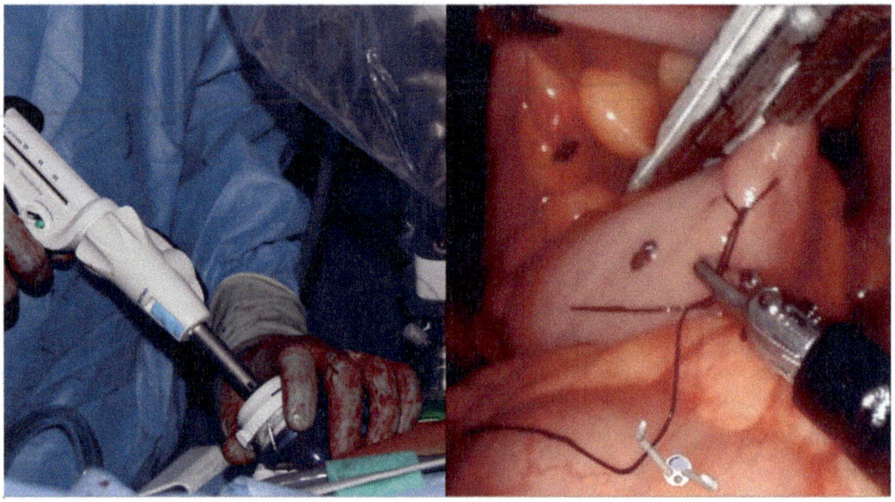

FIGURE 7.4

Intracorporeal bowel harvesting prior to perform the urinary diversion in male. The assistant surgeon utilizes the 12 mm Airseal port to introduce the EndoGia bowel stapler.

7.4 Ileal conduit

Incontinent diversion with using ileum to direct urine from a reservoir to an external collecting bag via a stoma is one of the most common techniques for urinary diversion. Similar bowel harvesting techniques are used as described earlier.

After performing the radical cystectomy, attention is turned to create the ileal conduit. The appendix is identified and then the terminal ileum. From there, 15 cm are measured proximally. A 3–0 Vicryl stitch is placed to mark the distal portion of the conduit. A 3–0 Silk stitch is placed 15 cm proximal to that to mark out the proximal portion of the conduit (Fig. 7.5).

At these two areas, the mesentery is isolated away from the ileum. The mesentery is then taken with electrocautery, leaving about an inch and a half of mesentery proximally and distally. The proximal (Fig. 7.6) and the distal (Fig. 7.7) portions of the future conduit are isolated from the rest of the bowel by using a stapler.

Once this step is completed, reanastomosis of the bowel is then performed as mentioned earlier.

An enterotomy is then made at the distal and proximal stumps of the ileum (Fig. 7.8).

These two ends are brought together with a Vicryl stitch, then the stapler is fired to establish continuity between the two segments (Fig. 7.9).

One load is used to establish the continuity between each segment; another load is then used to close off that segment to complete the bowel reanastomosis.

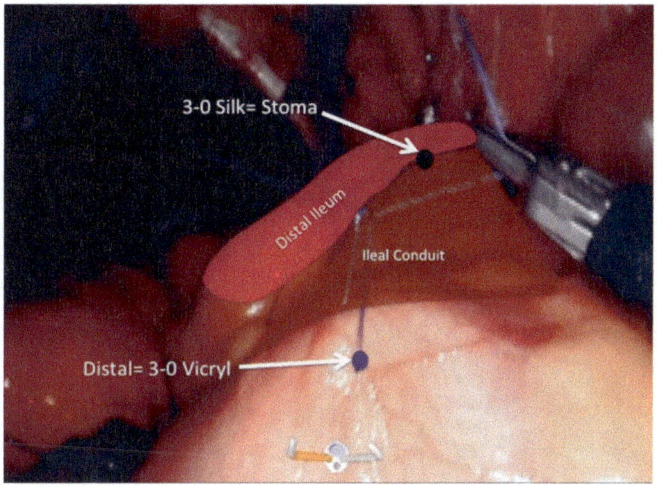

FIGURE 7.5

Identification and marking of the ileal conduit. The distal and the proximal ends of the small bowel segment are marked with 3–0 silk and 3–0 dyed polysorb stay sutures.

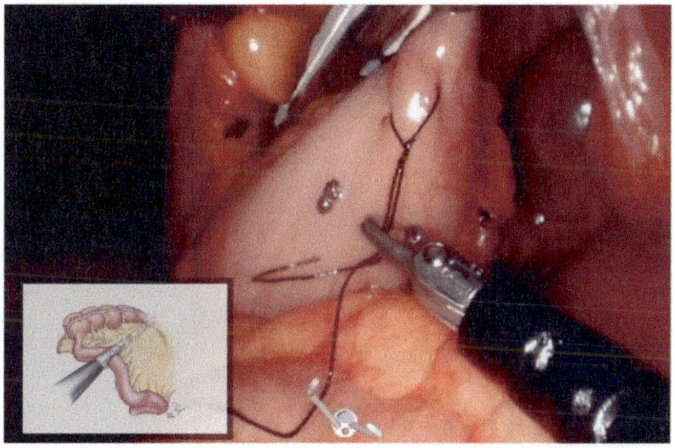

FIGURE 7.6

Proximal segment of the ileal conduit. The ileum segment is transected on either side by using a 45-mm laparoscopic stapler maneuvered through the assistant port.

Attention at this point is made to the ureteroileal anastomosis. The left ureter is brought under the sigmoid mesentery to the right side with measures made to avoid tension. Both the right and the left ureters are spatulated (Fig. 7.10) and opened to perform a Wallace or Bricker (according to the surgeon's preference) ureteroileal anastomosis (Fig. 7.11).

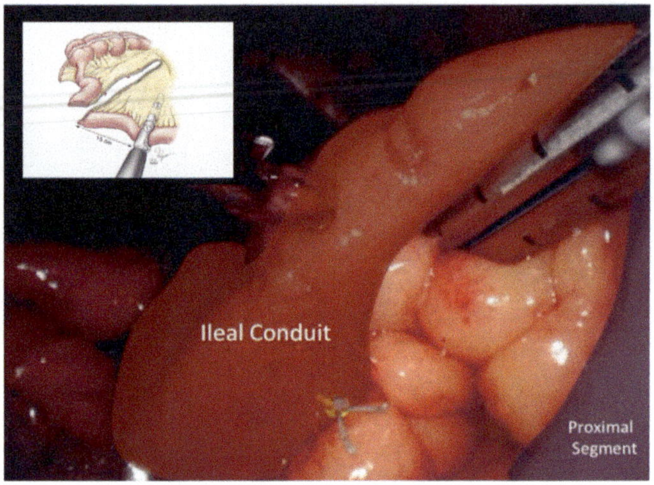

FIGURE 7.7

Distal segment of the ileal conduit. Same as for the proximal segment, the conduit is transected on either side by using a 45-mm laparoscopic stapler.

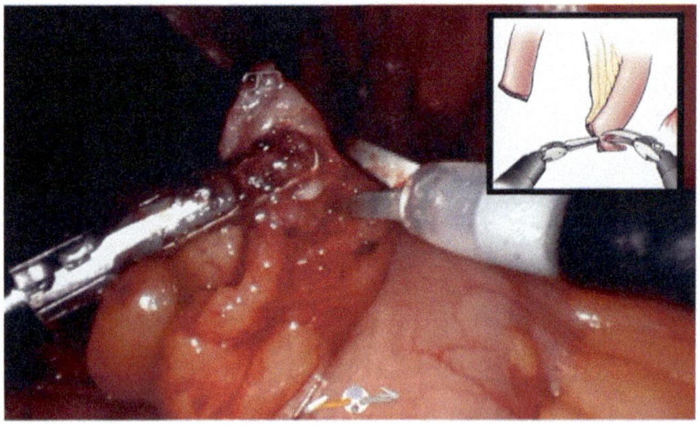

FIGURE 7.8

Two small enterotomies are created proximally to the staple lines for the insertion of the laparoscopic stapler.

The medial walls of each ureter are anastomosed together with a 4–0 Monocryl. Then an enterotomy is made in the proximal portion of the future conduit. Another 4–0 Monocryl is used to bring together the combined ureteral segment with the enterotomy, starting with a stitch at the apex of the enterotomy and

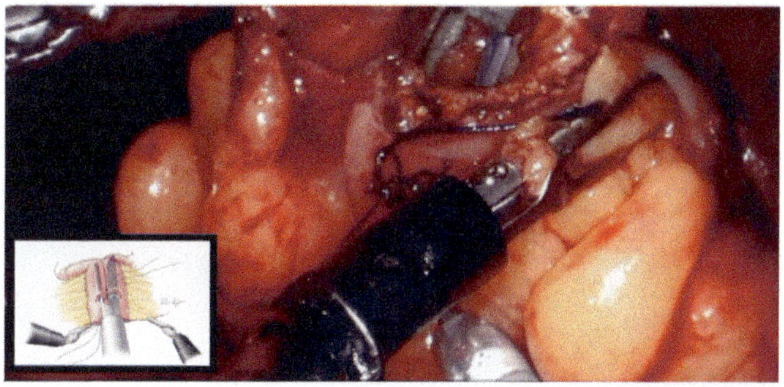

FIGURE 7.9

Through the assistant port, a laparoscopic stapler with 45-mm bowel loads is inserted into the bowel segments to perform a side-to-side bowel anastomosis.

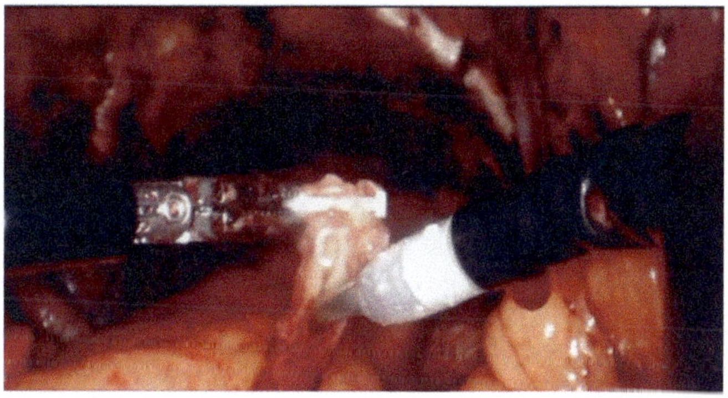

FIGURE 7.10

Each ureter is spatulated in preparation for the ureteroenteric anastomosis.

of the combined ureteral segments. Two ureteral catheters are placed in each ureter before the anastomosis is completed.

At this point, with the conduit ready to be matured through the stoma, the robot is undocked; the distal ileum segment is brought up by using a grasper. The skin around the port site is incised circumferentially down to the level of the fascia. The conduit is then brought through the wound.

A 3–0 Vicryl to tack the serosa of a more proximal segment of the conduit to the fascia at the 3, 6, and 9 o'clock positions, avoiding the mesentery at the 12 o'clock position. The authors then place 3–0 Vicryl stitches circumferentially to evert the bowel mucosa and tack it to the skin, forming a rosebud stoma.

Developing the stoma is critical in avoidance of future complications. In the authors' experience, the rate of complications was reduced involving the stoma by following the literature recommendations for stoma creation.[12] Stoma extends

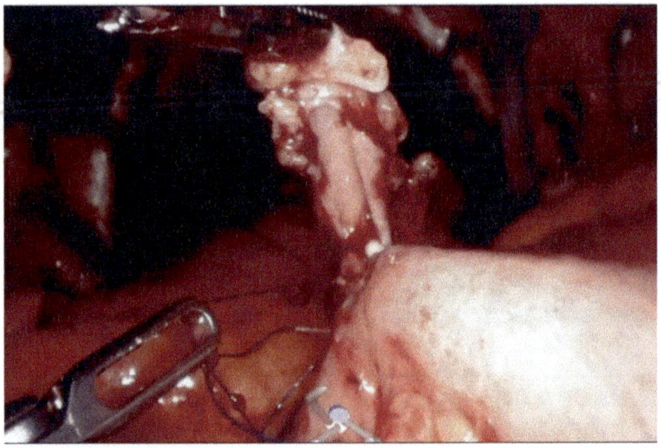

FIGURE 7.11

Separate ureteroenteric anastomoses are performed according to the Bricker technique by using two 4–0 Polysorb running sutures.

one to two inches above the edge of the skin to reduce the risk of stenosis and providing a better fit for the stoma appliance. In addition, the stoma extends through the belly of the rectus muscle as extending it lateral to the muscle could lead to increased risk of parastomal hernia.

7.5 Neobladder

Studer described neobladder creation in which the bowel is folded on itself in a U shape leaving a chimney.[13] At the authors' institution, the Studer neobladder techniques are commonly adopted. Using the single port, the apex of the bowel segment is anastomosed to the posterior urethra. The ileum is then detubularized to make the pouch spherical for the neobladder compliance purposes. This is created using 3–0 barbed sutures. At this point, closing the anterior urethral part completes the anastomosis. The distal ends of the ureters are then spatulated medially and sutured to form a single conjoined free end that is then anastomosed to the chimney part of the pouch, using the Wallace technique. Double J stents are then placed and a 22 Fr silicone catheter is left in place through the urethra and irrigated with 150 mL of normal saline solution to test for any eventual leakage. When creating a neobladder, the dissection is critical for the postoperative recovery of continence. In males, the apical dissection to keep the striated muscle intact as well as the apex of the prostate helps in preserving this. In the authors' common practice, antegrade and retrograde nerve sparing technique is performed to help with the recovery of sexual function postoperatively. The urethral length makes the anastomosis easier. When looking at preserving females postoperative continence, there are also important steps. First one is to keep the anterior vagina below the urethra and trigone by only removing the anterior and posterior fornix. This allows for enough space to use instruments for stapling. Second, if the patient would like to preserve the uterus, then only opening the anterior fornix would allow enough space for delivery of the specimen.

7.6 Indiana pouch

When creating an Indiana pouch with a single-port robotic system, there are important benefits with some small differences. The Indiana pouch operation is performed in conjunction with a radical cystectomy. The Indiana pouch needs the ileocecal valve as its continence mechanism. This operation is performed by taking 60 cm of ascending colon and transverse colon to create the pouch and 12 cm of the terminal ileum to bring up a stoma. Typically, in a robotic multiarms, multiports case, the robot is first docked close to the umbilicus and directed down into the pelvis. Once the radical cystectomy is finished, the cecum and the ascending colon are then mobilized by dissection along the line of Toldt. The same steps are reproduced when using the single port robot.

A key benefit of the single-port robot is that some necessary steps encountered when using a multiarms robotic platforms are avoided, namely the undocking of the robot, the patient repositioning, and the redocking. With the single port, the robot has the ability to reposition intracorporeally to continue the case, thanks to the rotating boom. The mobilization of the ascending colon continues until 60 cm is reached which requires dissection of the hepatocolic ligament, past the hepatic flexure to the transverse colon. The colon is mobilized enough to start the creation of the Indiana pouch.

Starting the Indiana pouch can be done either intracorporeally or extracorporeally. In single port, the authors' practice starts with an extracorporeal approach; however, it can technically be done entirely intracorporeally. The bowel anastomosis can be done like the neobladder anastomosis except for the small bowel, a single 60 cm staple load is used to create the two free ends and with the large bowel, two staple loads are used to create a bigger anastomosis. With the removal of the ileocecal valve, the authors did not observe a difference in long-term quality of life.

Next to create the pouch, the colon is dissected along the transmesenteric border and then folded over to detubularize the bowel. The two ends of the pouch are sutured together with running 3–0 barbed Vicryl suture. After this is created, the stoma needs to be fashioned to be catheterizable. The small bowel and the ileocecal valve create this channel. The small bowel is imbricated by placing a foley catheter in the small bowel and using multiple staple loads to come across the bowel until reaching the cecum to decrease its size. A 3–0 Vicryl is then used to constrict the tissue further to ensure the long-term adequacy.

Lastly, the ureteral anastomosis has to be completed. This is done with a Bricker technique anastomosis of the ureter to the end of the cecum. A 4–0 Monocryl is used for the ureteral anastomosis with a double-J stent placed in the ureter and pouch. To perform a nonrefluxing anastomosis, it would need to be tunneled. A 20 Fr Malecot catheter is placed in the pouch and then through the skin. This catheter is used for daily irrigation. This large catheter allows for minimal mucus to clog the catheter. A 14 Fr catheter will be placed through the stoma and left capped to allow the stoma to heal and as a backup drainage system.

Lastly, the maturing of the stoma takes place. The single-port incision at the umbilicus is a guide as where to make the stoma. This initial incision is 2 cm from the umbilicus to provide ample space for the stoma. The stoma is matured in similar manner to the ileal conduit as mentioned earlier. The single-port incision is capped. The Malecot catheter is passed through a separate skin incision away from the stoma and capped.

In the authors' practice, it was found that the Malecot drainage system and the stoma catheter provide better patient experience and function. The double-J stent allows for the stent to be secured and a urostomy appliance is placed over the stoma. The stents are removed at 4 weeks postoperatively via cystoscopy. The Malecot catheter is removed after 4 weeks and the next day the stoma catheter is removed because this extra day helps to close the opening in the pouch that was utilized for the Malecot catheter.

7.7 Delivering the specimen

EndoCatch device is used to remove the specimen. There is a difference in delivering the specimen between males and females. In males, the single-port incision is extended to deliver the specimen, while in females the natural vaginal orifice can be used to deliver the specimen. Given that the vaginal orifice is used as an entry port for the stapler, this natural orifice can be closed with a 3–0 barbed suture after delivery of the specimen and the purpose of this orifice is achieved.

7.8 Benefits and disadvantages of the single-port technique

One of the most important reasons why the authors nowadays tend to use the single-port system for performing urinary diversions is the administration of less narcotic medications postoperatively. The data observed in the authors' experience significantly showed that single-port patients used nearly 14 mg fewer of morphine equivalents during their hospitalization compared to patients who undergo surgery performed by using multiarms robotic systems. Given that less narcotics correlate with earlier return of bowel function, this will also help with postoperatively recovery.

A key tool in single-port surgery is the navigator in the screen of the single-port system providing a real temporal and spatial positioning of the robotic arms while operating that could help the surgeon with avoiding collisions and knowing where arms are when not visible on the screen. The single-port robot has two elbows and the proximal elbow is often off screen. The navigator allows the surgeon to be fully aware of the instruments' positioning regardless of what is visible on screen. In addition, single-port technology also provides better cosmesis due to the single surgical skin incision to be closed compared with multiple incisions common after surgery performed with multiarms robots. Patients have given more satisfactory comments in this regard (Fig. 7.12).

Beyond the advances that the adoption of the single-port robotic platform does allow, one should acknowledge there can also be significant challenges. These include the ergonomic challenges of inhabiting a large extraperitoneal space; however, this has not impacted on the operative time in the authors' experience. There are technical learning curves as well. For example, wide movements with the instruments of the single-port robotic platform are prevented: particularly, the capability of traction with the fourth arm is reduced. Moreover, all the four arms have to be moved in tandem. The difficulty with moving arms in initial experiences has been reported with increased rate of clashing of robotic elbows. This can cause rapid and potentially dangerous arm movements. Another challenge to be kept in mind is the clashing of instruments especially at the proximal elbow, which is commonly out of the surgeon's view. There is the need to be

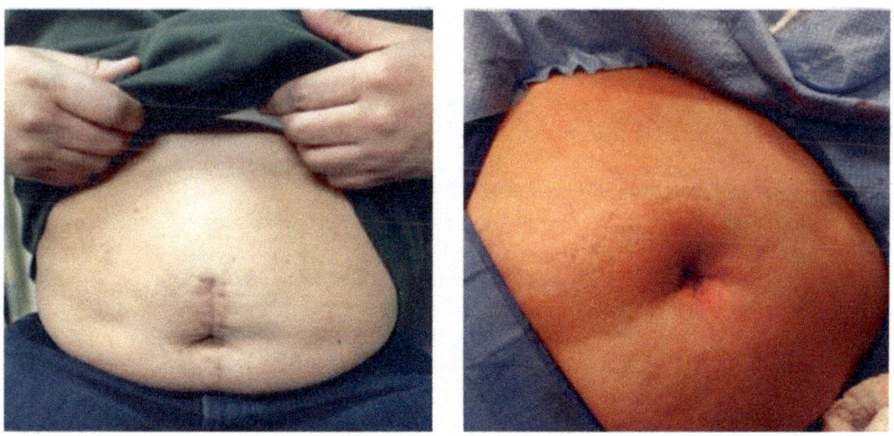

FIGURE 7.12

Cosmetic results of single-post incisions following neobladder surgeries with image in the left showing incision above and including the umbilicus and image on the right showing incision through the umbilicus.

"hyperaware" about the position of the robotic instruments placement during radical cystectomy: this is made possible by looking to the navigator, available with the SP platform as mentioned earlier.

Another challenge to mention when using the single-port system is suturing. Tying knots down laterally is very difficult given the range of movement of the single-port system; it creates a challenge that can be overcome with experience. Efficient knots tying with the single-port robot utilizes the z-axis of movement. As such, the single-port robot excels at moving instruments in and out but it is much more limited in lateral movements in the x- and y-axis.

To use the ostomy site as a port for suction and traction can be useful. If it is a neobladder case, then a port in the left lateral abdominal wall is introduced if needed. In women, the vaginal cuff can also be a point of entry for suctioning or stapling, as described.

In the authors' experience, no significant differences were found in complication rates between multiport and single port for performing urinary diversion. The operation performed with the SP robot mimics that of the multiarms, multiport robot with some technical differences and the relative learning curve. With proper training and high volume of experience, single-port urinary diversion can be performed effectively.

References

1. Pannek J, Senge T. History of urinary diversion [Internet]. *Urol Int.* 1998;60:1–10 [cited 2021 Jun 9]. Available from: https://pubmed.ncbi.nlm.nih.gov/9519414/.
2. Menon M, Hemal AK, Tewari A, et al. Nerve-sparing robot-assisted radical cystoprostatectomy and urinary diversion. *BJU Int [Internet].* 2003;92(3):232–236. Available from: https://pubmed.ncbi.nlm.nih.gov/12887473/.
3. Parekh DJ, Reis IM, Castle EP, et al. Robot-assisted radical cystectomy vs open radical cystectomy in patients with bladder cancer (RAZOR): an open-label, randomised, phase 3, noninferiority trial. *Lancet [Internet].* 2018;391(10139):2525–2536. Available from: https://pubmed.ncbi.nlm.nih.gov/29976469/.
4. Venkatramani V, Reis IM, Castle EP, et al. Predictors of recurrence, and progression-free and overall survival following open vs robotic radical cystectomy: analysis from the RAZOR trial with a 3-year follow-up. *J Urol [Internet].* 2020;203(3):522–529. Available from: https://pubmed.ncbi.nlm.nih.gov/31549935/.
5. Tanneru K, Jazayeri SB, Kumar J, et al. Intracorporeal vs extracorporeal urinary diversion following robot-assisted radical cystectomy: a meta-analysis, cumulative analysis, and systematic review [Internet]. *J Robotic Surg.* 2021;15:321–333 [cited 2021 Jun 9]. Available from: https://pubmed.ncbi.nlm.nih.gov/33222043/.
6 Hussein AA, Elsayed AS, Aldhaam NA, et al. A comparative propensity score-matched analysis of perioperative outcomes of intracorporeal vs extracorporeal urinary diversion after robot-assisted radical cystectomy: results from the International Robotic Cystectomy Consortium. *BJU Int [Internet].* 2020;126(2):265–272 [cited 2021 Jun 9]. Available from: https://pubmed.ncbi.nlm.nih.gov/32306494/.

7. Zhang JH, Ericson KJ, Thomas LJ, et al. Large single institution comparison of perioperative outcomes and complications of open radical cystectomy, intracorporeal robot-assisted radical cystectomy and robotic extracorporeal approach. *J Urol [Internet]*. 2020;203(3):512–521 [cited 2020 Aug 23]. Available from: https://pubmed.ncbi.nlm.nih.gov/31580189/.
8. Kaouk J, Garisto J, Eltemamy M, Bertolo R. Single-port robotic intracorporeal ileal conduit urinary diversion during radical cystectomy using the SP surgical system: step-by-step technique. *Urology*. 2019;130:196–200. Available from: https://pubmed.ncbi.nlm.nih.gov/30965039/.
9. Zhang M, Thomas D, Salama G, Ahmed M. Single port robotic radical cystectomy with intracorporeal urinary diversion: a case series and review. *Transl Androl Urol [Internet]*. 2020;9(2):925–930 [cited 2020 Nov 8]. Available from: http://tau.amegroups.com/article/view/37475/30269.
10. Billah MS, Stifelman M, Munver R, Tsui J, Lovallo G, Ahmed M. Single port robotic assisted reconstructive urologic surgery-with the da Vinci SP surgical system. *Transl Androl Urol [Internet]*. 2020;9(2):870–878 [cited 2020 Aug 23]. Available from: https://pubmed.ncbi.nlm.nih.gov/32420202/.
11. Ali D, Sawhney R, Billah MS, et al. Single-port robotic radical cystectomy with intracorporeal bowel diversion: initial experience and review of surgical outcomes. *J Endourol*. 2022;36:216–223.
12. Madersbacher S, Schmidt J, Eberle JM, et al. Long-term outcome of ileal conduit diversion. *J Urol*. 2003;169(3):985–990.
13. Studer UE, Ackermann D, Casanova GA, Zingg EJ. Three years' experience with an ileal low pressure bladder substitute. *Br J Urol [Internet]*. 1989;63(1):43–52 [cited 2020 Oct 18]. Available from: https://pubmed.ncbi.nlm.nih.gov/2920259/.

CHAPTER

Single-port robotic ureteral reconstruction

8

Nathan Cheng and Michael Stifelman

Department of Urology, Hackensack Meridian School of Medicine, Nutley, NJ, United States

8.1 Introduction

Prior to the da Vinci SP (Intuitive Surgical Inc.) surgical system's introduction in 2018 and subsequent popularization in urologic robotic surgery, laparoendoscopic single-site surgery (LESS) was a technique that was introduced a decade prior to minimize port-related complications, decrease postoperative pain, and improve cosmesis (Fig. 8.1).[1] Despite its feasibility in urologic surgery to achieve those outcomes,[2] LESS was not widely adopted due to its steep learning curve as well as the concurrent popularization of robotic-assisted laparoscopy. In particular, ureteral reconstruction is a suturing-heavy subset of urologic surgery, in which the ergonomics and wrist articulation advantages of robotic-assisted surgery is superior to that of laparoscopic technique. As the da Vinci SP system came out, the

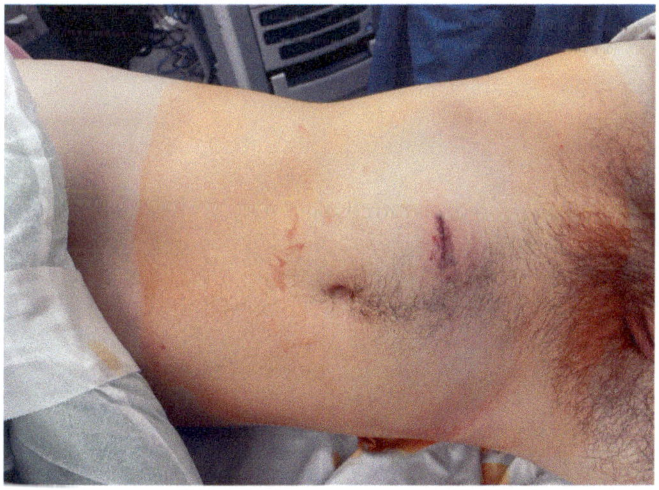

FIGURE 8.1

Example of a left lower quadrant single-port incision for a robotic right pyeloplasty on a young patient.

Single-Port Robotic Surgery in Urology. DOI: https://doi.org/10.1016/B978-0-323-91906-7.00001-6

advantages of LESS were combined with the surgeon ergonomics of multiport da Vinci robots.

8.2 Advantages

The single port (SP) system allows for dynamic external range of motion, allowing the surgeon to change the target anatomy without redocking the robot or changing trocar position. This proves very useful when operating on organs far apart in the body, for example, in nephroureterectomy, where the surgeon can efficiently move from the kidney down the ureter to the bladder while not having to reposition the trocar (although undocking and putting the patient in Trendelenburg is helpful) and without the issue of clashing working arms. The majority of ureteral reconstructions (i.e., pyeloplasty, ureteroureterostomy, ureteral reimplantation) are surgeries on small areas in the body, so dynamic retargeting does not need to be utilized too often. Despite not needing to utilize this advantage because of the usually small area of target anatomy, the SP robot is advantageous in operating in tight regions that would otherwise be difficult for multiple arms to get into. One example of this is performing ureteral reconstruction in the retroperitoneal space, where it may be more difficult to place five trocars into limited mid-clavicular or flank territory, depending on body habitus. Another example of a region that may be accessed easier with the single-port robot compared to multiport is deep pelvis surgery for bladder neck or proximal urethral reconstructions.

8.3 Disadvantages

This first iteration of the da Vinci SP robotic system is not without its weaknesses.

A minimum distance of 10 cm (although upto 25 cm is ideal) between the end of the cannula trocar and the target organ should be met in order to allow for full deployment of the elbow and wrist angulation capabilities to minimize collisions during operation; this is generally not an issue when utilizing the SP Access Port Kit (Intuitive Surgical Inc.) (Fig. 8.2), but selection of the initial incision must account for this minimum distance for surgeons who do not have the SP Access Port Kit and choose to utilize the Alexis wound protector with GelPort (Applied Medical, Rancho Santa Margarita, CA).[3]

While there is no published study to date addressing the learning curve associated with the da Vinci SP robotic system, there have been multiple studies addressing a steeper learning curve in other single-site minimally invasive techniques across multiple surgical subspecialties.[4–6] The learning curve with the da Vinci SP may be attributed to multiple factors, including less freedom with retraction,

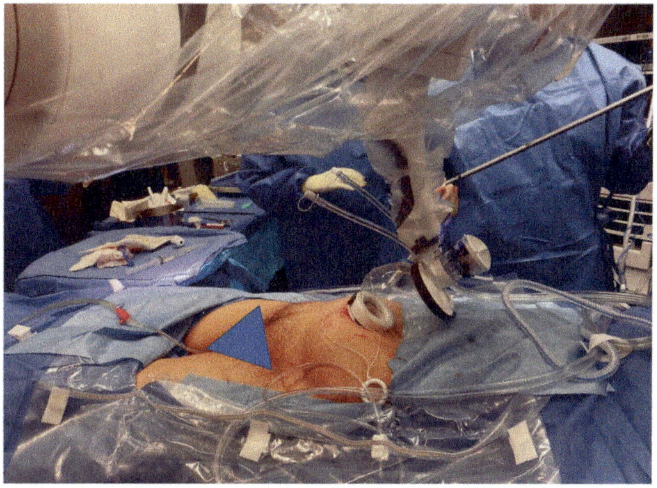

FIGURE 8.2

SP Access Port Kit (Intuitive Surgical Inc., Sunnyvale, California, United States) allowing for additional trocar to target organ distance due to its "air-docking."

utilization of "elbow" instead of "wrist" articulation, and lack of certain instrumentation. With regards to retraction, the limitations of the da Vinci SP arms reside in the inability to move the arm too far away from the camera and other instruments—thus, retraction must be performed on a smaller scale with less tissue involved. An example of this is putting the renal hilum on stretch during a nephrectomy or partial nephrectomy: in the da Vinci Xi multiport system, the retracting arm can lift up between the psoas and the entire kidney; however, in the da Vinci SP single port system, the retracting arm would need to stay closer to the hilum so that the working arms can dissect out the vessels. Suturing with the da Vinci SP needle drivers is also slightly different from the da Vinci Xi and other previous multiport iterations, as there is somewhat less wrist articulation and more elbow articulation. As the da Vinci SP is in its first iteration, there are some instruments that have not yet been available to utilize, such as suture-cut needle drivers and vessel-sealing devices.

A popular tool in robotic ureteral reconstruction has been the use of indocyanine green (ICG) and Firefly fluorescence imaging (Intuitive Inc.), which is near-infrared technology that allows the surgeon to visualize the perfusions of tissue when ICG is administered intravenously or to visualize the ureter when surrounded by thick fibrosis when ICG is administered intraureterally. This first iteration of da Vinci SP robot does not come with Firefly technology.

Lastly, just how there is a learning curve with the console surgeon, there is also a learning curve with the bedside assistant. Some surgeons utilize the single-port "plus one" model, which is the single-site da Vinci SP access in addition to an assistant trocar through another incision, as it gives the bedside assistant more angles to retract when necessary and freedom from clashing with the robotic arms. The true single-port approach without a "plus one" involves the assistant port through the same single-incision; the proximity of the assistant port to the da Vinci SP port and working arms means that the bedside assistants have a limited radius that they may be able to assist and that clashing with the robotic arms is a frequent occurrence. This may make passing sutures, bulldog clamps, clips, and staples more difficult.

8.4 Access

The new SP Access Port Kit (Intuitive Inc.) serves to mitigate some of the limitations that come with the da Vinci SP system, including the minimum trocar to target organ distance and bedside assistance. It comes with the wound protector, as well as the clear spherical attachment which has three separate ports—one for the da Vinci SP trocar, one for 12 mm instrumentation, and one for an 8 mm assistant port on the side of the sphere. An Airseal (CONMED Corporation, Utica, NY) 8 mm trocar is typically used for the side port (Fig. 8.3).

After docking the robot, there are some differences between the SP and multiport systems that the surgeon at the console will notice. A new schematic, called the Navigator, is at the bottom of the screen, mapping out the positions of each

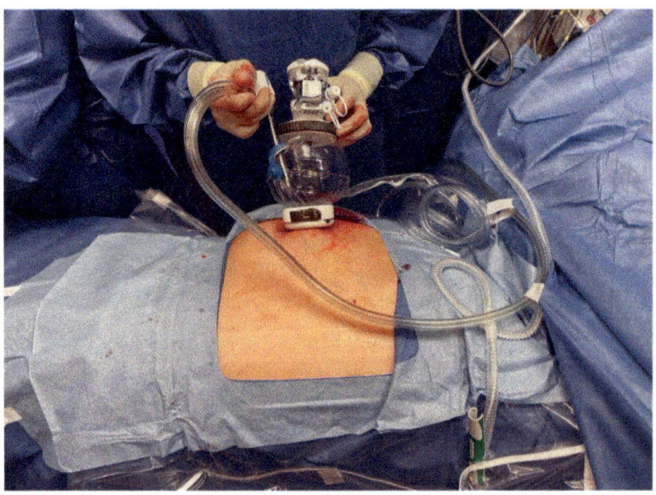

FIGURE 8.3

SP Access Port Kit (Intuitive Surgical Inc., Sunnyvale, California, United States) with 8 mm Airseal trocar (CONMED Corporation, Utica, New York, United States) through the side port for insufflation.

instrument in relation to each other, allowing the surgeon to identify the location of the instruments when not in view as well as troubleshooting instrument collisions. The Navigator will also show the angulation of the camera, which is the first fully elbowed 3D HD camera that can be bent into "cobra" configuration to articulate into 30 degree down.

The remotely operative suction irrigation system (ROSI) is very useful in SP robotic surgery. It is a flexible tubing with a short rigid tip that the surgeon is able to grasp and maneuver throughout the operative field. It is particularly useful in SP robotic surgery, as it is less ergonomic for the bedside assistant to utilize a rigid laparoscopic suction device without clashing with the robotic instruments in true single-site surgery. There are few circumstances where an additional assistant port may be placed into the abdomen that would allow the bedside assistant to be more active in the surgery, for example, when there needs to be additional retraction or more active suctioning.

8.5 Pyeloplasty

Pyeloplasty was one of the first urologic robotic surgeries, not just specific to ureteral reconstruction, to be performed utilizing the single-port robot. Ureteropelvic junction (UPJ) obstruction is a condition often diagnosed in younger patients, a population who seeks improved cosmesis. It is also a surgery with a small operative field, making the SP robotic system well-suited to complete the task effectively.

Patients are positioned in lateral decubitus, and the genitalia is often prepped into the field to be able to perform cystoscopy postureteral stent placement to confirm placement. Some surgeons do not perform cystoscopy to confirm the distal curl of the ureteral stent under direct visualization.

The SP robotic port placement can be variable for pyeloplasties, but usually are either periumbilical or "mini-Pfannenstiel," as described by Kaouk et al.,[7] which is more accurately detailed as a transverse incision along Langer's lines at the ipsilateral lower quadrant similar to that of a Gibson incision.

Intra-operatively, the surgical technique is similar to that of the multiport approach, where the UPJ is identified by reflecting the colon medially in most cases, and transmesenteric approach used in select pediatric cases. The literature on SP robotic pyeloplasty[6–8] describe dismembered technique, excising the stenotic UPJ segment, spatulating the proximal ureter, and anastomosing the ureter to the renal pelvis with 4–0 absorbable suture over an indwelling ureteral stent. In the case of crossing vessels, the anastomosis is transposed anterior to the vessel (Fig. 8.4).

While still relatively in its infancy, SP pyeloplasty has been shown to have promising outcomes compared to the standard multiport robotic approach, with

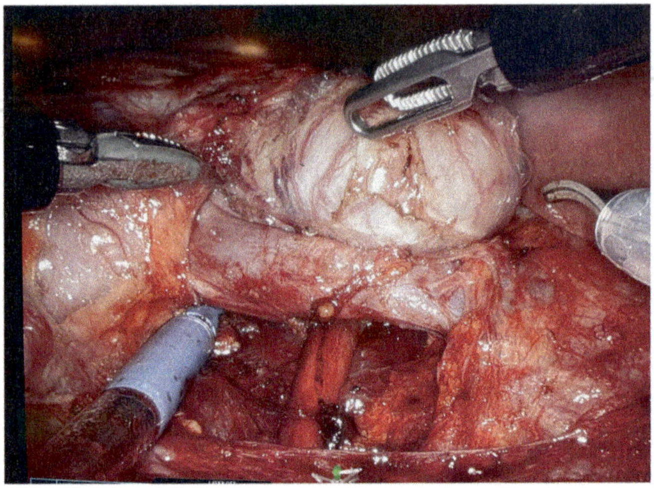

FIGURE 8.4

Robotic right pyeloplasty utilizing the da Vinci SP robot (Intuitive Surgical Inc, Sunnyvale, California, United States). Ureteropelvic junction (UPJ) obstruction secondary to crossing vessel: large accessory vein with smaller accessory artery overlying the UPJ with resultant hydronephrosis. Kidney, left; inferior vena cava, right; renal pelvis, top; ureter, bottom.

improved pain scores at discharge, lower opioid usage, and potential for same-day discharge.[8]

8.6 Ureteroureterostomy and buccal mucosa graft ureteroplasty

Ureteroureterostomy is typically utilized for proximal or mid ureteral strictures that are short enough to excise and primarily suture together without tension, traditionally described to be <3 cm in length.[9] Although it is early in the da Vinci SP experience, the surgical steps in utilizing the SP robot for primary ureteroureterostomy are identical to that of multi-port robotic ureteroureterostomy: identifying and dissecting out the strictured ureteral segment, transecting the ureter distally and proximally where there is healthy tissue, spatulating, and performing anastomosis around a ureteral stent.

There have been early experiences with more complex variations of proximal or mid-ureteral stricture repairs reported in the literature. Buccal mucosal graft ureteroplasty has been shown to be safe and feasible in proximal or mid-ureteral strictures too long for primary ureteroureterostomy.[3]

Buccal mucosal graft ureteroplasty has been a novel technique that extrapolated concepts from lower tract urethral reconstruction in utilizing tissue from the inner cheek to incorporate into the urinary system, as buccal mucosa is easy to harvest with low morbidity, compatible with wet environment, lacks hair, and has a highly vascular lamina propria that allows for successful graft assuming a well-vascularized environment. There have been multiple studies on multiport robotic buccal mucosal graft ureteroplasties demonstrating high success and low complication rates.[10]

The beginning of the buccal mucosal graft ureteroplasty is similar to that of a primary ureteroureterostomy, where the diseased ureteral segment is first identified and dissected out. At this point, it is important to know whether the ureteral lumen is narrowed or obliterated—if the lumen is narrowed, then a onlay buccal mucosa graft ureteroplasty is performed; however, if the lumen is obliterated, then an augmented anastomotic buccal mucosal graft ureteroplasty may need to be performed. In the onlay procedure, an ureterotomy is performed over the anterior aspect of the strictured segment so that apices of the ureterotomy are at healthy ureter (Fig. 8.5). The buccal mucosa graft is then sutured over the ureterotomy with absorbable suture over an indwelling ureteral stent (Fig. 8.6). In the augmented anastomotic procedure, the obliterated segment is excised, and the posterior aspects of the proximal and distal segments are anastomosed together to create a posterior plate so that the anterior aspect of the ureter is open and can be treated like the onlay buccal mucosal graft ureteroplasty. It is important to understand that while this has been shown to be a successful technique, the posterior

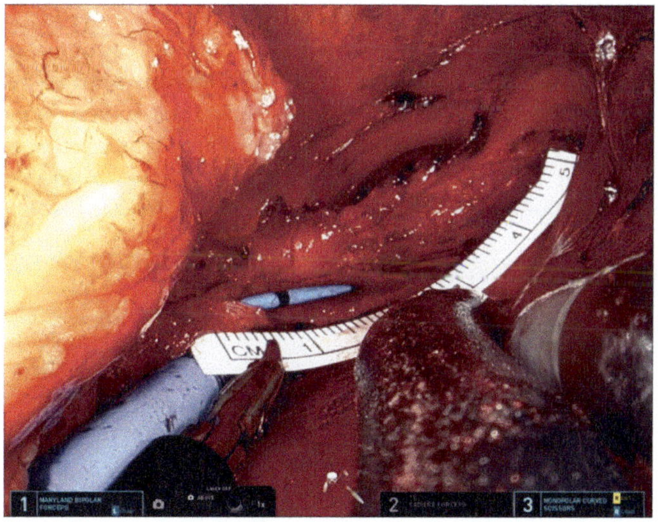

FIGURE 8.5

Retroperitoneal onlay buccal mucosa graft ureteroplasty utilizing the da Vinci SP robot (Intuitive Surgical Inc, Sunnyvale, California, United States). Anterior ureterotomy over a 5-french ureteral catheter.

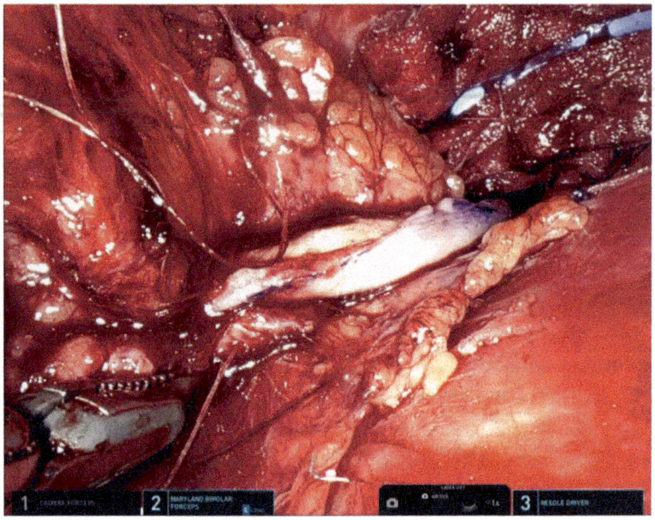

FIGURE 8.6

Retroperitoneal onlay buccal mucosa graft ureteroplasty utilizing the da Vinci SP robot (Intuitive Surgical Inc, Sunnyvale, California, United States). Suturing the buccal mucosal graft with shiny mucosa side intraluminal and dull lamina propria side extraluminal.

ureteral plate may not be possible to create if the obliterated segment excised was too long.

After the anastomosis of the buccal mucosal graft to the ureter, it is important to wrap the ureter in well-vascularized tissue to allow the graft to take. There are multiple options described, most commonly omentum, although mesentery and perinephric fat have also been used (Fig. 8.7). When creating the wrap of tissue circumferentially, it is important to ensure that the outside of the buccal mucosa graft, which is its lamina propria, has maximal contact with the vascularized flap.

8.7 Ureteral reimplantation

Ureteroneocystostomy, or ureteral reimplantation, is a tool used to treat ureteral strictures that are distal enough to be able to treat by reimplanting into the bladder, with or without mobility adjunctive maneuvers such as psoas hitch or Boari flap. Robotic-assisted techniques for ureteral reimplantation have been well-described to be feasible and comparable to their laparoscopic or open counterparts.[11,12] The single-port robotic technique for ureteral reimplantation was quickly adopted and described initially in a case series of three patients,[13] not long after to be part of larger data sets and studies. The initial case series utilized the GelPOINT access platform (Applied Medical, Rancho Santa Margarita,

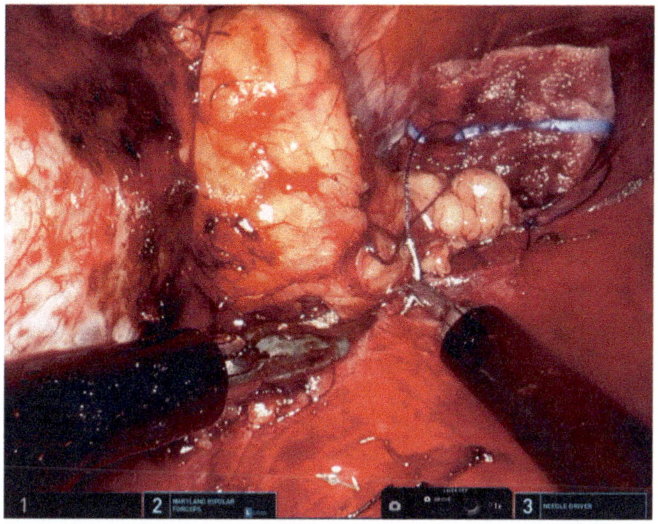

FIGURE 8.7

Retroperitoneal onlay buccal mucosa graft ureteroplasty utilizing the da Vinci SP robot (Intuitive Surgical Inc, Sunnyvale, California, United States). Perinephric fat wrap around the buccal mucosal graft ureteroplasty.

California, United States) for the SP trocar as well as a 12-mm bedside assistant port to go through. Now, most of the access for the robotic trocar is via the SP Access Port Kit (Intuitive Inc, Sunnyvale, California, United States), where there is a side port to accommodate an 8 mm trocar, as well as a port at the top next to the robotic trocar, which accommodates 12 mm instruments.

Patient position is supine with moderate to steep Trendelenburg angulation. In female patients modified lithotomy or "frogleg" positioning is often required for urethral access. The port positioning is usually supraumbilical, but depending on the location of stricture, can be moved more cephalad or caudal. After the strictured segment of ureter has been identified, it is traced up proximally until healthy ureter to be reimplanted into the ureter is seen. Classically, the ureter is transected at this point, although a new technique for nontransecting side-to-side ureteral reimplantation has been introduced,[14] which will be discussed later. The bladder is then filled, and it is determined visually if there is enough capacity to meet the ureter in a tension-free manner. If not, adjunctive measures such as a psoas hitch or Boari flap must be performed. The ureter is then spatulated approximately 2.5 cm and anastomosed to a cystotomy of the same size with absorbable suture. Mucosa-to-mucosa apposition is of utmost importance. Prior to completion of the anastomosis, an indwelling ureteral stent is placed. An abdominal drain may be left up to the discretion of the surgeon, although advised if the ureteroneocystostomy anastomosis was tested to be nonwatertight. A Foley catheter is left for 7–10 days, and the indwelling ureteral stent around 4 weeks.

8.8 Psoas hitch

A psoas hitch is performed after dropping the bladder by developing the space of Retzius and creating enough laxity on the bladder by freeing up ipsilateral and, when necessary, contralateral to be able to tack it to the ipsilateral psoas fascia. Smooth or barbed absorbable sutures may be used, and on the psoas side, they should be thrown parallel to the muscle fibers in order to avoid entrapping the genitofemoral nerve.

8.9 Boari flap

Boari flap is a technique that should be used if there is not enough mobility on the bladder, even with a psoas hitch, to create a tension free anastomosis. The robotic technique for Boari flaps have been shown to have improved outcomes,[15] and although large single-port robotic studies have not yet been performed for Boari flaps, early series have been promising. Prior to reconstruction, the surgeon must confirm adequate bladder capacity to perform Boari flap. With the bladder filled between 300 and 500 mL, an inverted "U" is scored over the anterior surface of the bladder, making sure to keep the base of the flap wide in order to minimize risk for flap necrosis. The flap is then incised transmurally so that the apex of the flap is flipped cephalad to be anastomosed to the ureteral stump. The posterior aspect of the flap may also be tacked to the psoas fascia to further take off tension, if seen fit. The apex of the Boari flap is then anastomosed to the crotch of the posterior urethral spatulation with absorbable suture. The anastomosis is then taken circumferentially, and after completion, an indwelling ureteral stent is placed. The Boari flap is then tubularized with absorbable suture continuous with the cystotomy closure. This bladder closure may be done in either a one or two layer closure. The closure is then tested for watertightness in the usual fashion via Foley catheter. A surgical drain should be left. Because the bladder repair is more extensive than that of a regular ureteral reimplantation, the Foley catheter is often left for 10–14 days, and the ureteral stent removed in 4–6 weeks.

8.10 Nontransecting side-to-side ureteral reimplantation

The goal of a nontransecting side-to-side ureteral reimplantation is to avoid circumferential dissection of the ureter in order to minimize disruption of the adventitial vascular plexus. It is useful in patients with a severely fibrotic rind surrounding the strictured segment. Once the healthy ureter has been identified, the ureter is not transected. The ureter distal to the stricture is left in situ. A longitudinal ureterotomy approximately 3–4 cm are made over the anteromedial aspect of the healthy ureter, and this is anastomosed to the bladder

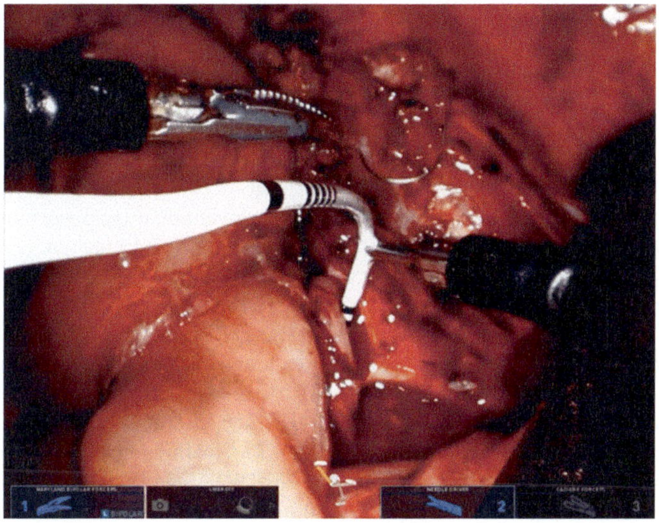

FIGURE 8.8

Left-to-right transureteroureterostomy utilizing the da Vinci SP robot (Intuitive Surgical Inc, Sunnyvale, California, United States). Cannulating the left ureter, which was brought submesenteric above the inferior mesenteric artery, with an indwelling ureteral stent.

in a side-to-side fashion. As a newly popularized technique, there have been no studies reporting single-port robotic outcomes; however, a recent multiinstitutional series of 16 patients has shown 100% radiologically improved hydronephrosis and 93.8% improvement in flank pain at median 12 month follow up.[14]

8.11 Summary

Robotic ureteral reconstruction is a continuously evolving field with recent advances including buccal mucosa ureteroplasty. With the popularization of the da Vinci SP robotic system, these ureteral reconstructions can be performed with the advantages of a single incision while being feasible and reproducible for the surgeon and safe and effective for the patient (Fig. 8.8).

References

1. Gettman MT, Box G, Averch T, et al. Consensus statement on natural orifice transluminal endoscopic surgery and single-incision laparoscopic surgery: heralding a new era in urology? *Eur Urol*. 2008;53:1117–1120.

2. Autorino R, Cadeddu JA, Desai MM, et al. Laparoendoscopic single-site and natural orifice transluminal endoscopic surgery in urology: a critical analysis of the literature. *Eur Urol*. 2011;59:26–45.
3. Billah MS, Stifelman M, Munver R, Tsui J, Lovallo G, Ahmed M. Single port robotic assisted reconstructive urologic surgery-with the da Vinci SP surgical system. *Transl Androl Urol*. 2020;9(2):870–878. Available from: https://doi.org/10.21037/tau.2020.01.06. PMID: 32420202; PMCID: PMC7214978.
4. Park P, Han HJ, Song TJ, et al. Single-port vs conventional laparoscopic distal pancreatectomy: a propensity score matched analysis and a learning curve of single-port approach. *J Hepatobiliary Pancreat Sci*. 2019;26(9):401–409. Available from: https://doi.org/10.1002/jhbp.646. Epub 2019 Jul 18. PMID: 31211913.
5. Chang CC, Yen YT, Lin CY, Chen YY, Huang WL, Tseng YL. Single-port video-assisted thoracoscopic surgery subsegmentectomy: the learning curve and initial outcome. *Asian J Surg*. 2020;43(5):625–632. Available from: https://doi.org/10.1016/j.asjsur.2019.09.009. Epub 2019 Oct 29. PMID: 31672480.
6. Kravetz AJ, Iddings D, Basson MD, Kia MA. The learning curve with single-port cholecystectomy. *JSLS*. 2009;13(3):332–336. PMID: 19793472; PMCID: PMC3015986.
7. Lenfant L, Wilson CA, Sawczyn G, Aminsharifi A, Kim S, Kaouk J. Single-port robot-assisted dismembered pyeloplasty with mini-pfannenstiel or peri-umbilical access: initial experience in a single center. *Urology*. 2020;143:147–152. Available from: https://doi.org/10.1016/j.urology.2020.05.041. Epub 2020 Jun 4. PMID: 32505618.
8. Beksac AT, Wilson CA, Lenfant L, et al. Single-port mini-pfannenstiel robotic pyeloplasty: establishing a non-narcotic pathway along with a same-day discharge protocol. *Urology*. 2021. Available from: https://doi.org/10.1016/j.urology.2021.10.013. S0090–4295(21)00986-9. Epub ahead of print. PMID: 34710396.
9. Andrade HS, Kaouk JH, Zargar H, et al. Robotic ureteroureterostomy for treatment of a proximal ureteric stricture. *Int Braz J Urol*. 2016;42:1041–1042.
10. Lee Z, Lee M, Koster H, et al. Collaborative of reconstructive robotic ureteral surgery (CORRUS). A multi-institutional experience with robotic ureteroplasty with buccal mucosa graft: an updated analysis of intermediate-term outcomes. *Urology*. 2021;147:306–310. Available from: https://doi.org/10.1016/j.urology.2020.08.003. Epub 2020 Aug 13. PMID: 32798516.
11. Skupin PA, Stoffel JT, Malaeb BS, Barboglio-Romo P, Ambani SN. Robotic vs open ureteroneocystostomy: is there a robotic benefit? *J Endourol*. 2020;34(10):1028–1032. Available from: https://doi.org/10.1089/end.2019.0715. Epub 2020 Jul 15. PMID: 32423300.
12. Buffi NM, Lughezzani G, Hurle R, et al. Robot-assisted surgery for benign ureteral strictures: experience and outcomes from four tertiary care institutions. *Eur Urol*. 2017;71:945–951.
13. Kaouk JH, Garisto J, Eltemamy M, Bertolo R. Robot-assisted surgery for benign distal ureteral strictures: step-by-step technique using the SP® surgical system. *BJU Int*. 2019;123(4):733–739. Available from: https://doi.org/10.1111/bju.14635. Epub 2019 Jan 4. PMID: 30499629.
14. Slawin J, Patel NH, Lee Z, et al. Ureteral reimplantation via robotic nontransecting side-to-side anastomosis for distal ureteral stricture. *J Endourol*. 2020;34(8):836–839.

Available from: https://doi.org/10.1089/end.2019.0877. Epub 2020 May 5. PMID: 32233674.
15. Do M, Kallidonis P, Qazi H, et al. Robot-assisted technique for boari flap ureteral reimplantation: is robot assistance beneficial? *J Endourol*. 2014;28:679–685.

Further reading

Steinberg RL, Johnson BA, Cadeddu JA. Ureteral reconstruction using the Da Vinci SP robotic platform: an initial case series. *J Endourol Case Rep*. 2019;5(2):60–63. Available from: https://doi.org/10.1089/cren.2018.0115. PMID: 32760802; PMCID: PMC7383444.

CHAPTER

9 Single-port robotic surgery for kidney transplantation and autotransplantation

Alireza Aminsharifi[1], Mahmoud Abou Zeinab[2] and Jihad Kaouk[2]

[1]*Department of Urology, Hershey Medical Center, Pennsylvania State University, Hershey, PA, United States*

[2]*Department of Urology, Glickman Urological and Kidney Institute, Cleveland Clinic, Cleveland, OH, United States*

9.1 Single-port robotic surgery for kidney transplantation—rationale and conceptualization

During the past decade, many centers adopted robotic-assisted kidney transplantation (RAKT) as an alternative minimally invasive procedure for kidney transplantation. A significantly higher rate of wound-related complications with open surgery has been the main driving force of the application of minimally invasive procedures for kidney transplantation. Large incisions used for wide exposure during conventional open kidney transplantation, particularly in the setting of immunosuppression, in patients with obesity diabetes, or other comorbidities are at a high risk of wound infection, dehiscence, and hernia. These complications would negatively affect patients' postoperative morbidity and recovery as well as the graft outcome.[1–7]

With the application of minimally invasive surgery for kidney transplantation, less wound complications and a quicker recovery could be expected. A systematic review and meta-analysis of data comparing minimally invasive versus standard open technique for kidney transplantation showed comparable graft and recipient survival outcomes with significantly less wound-related complications (0 vs 16%–19% vs 0%) with the application of minimally invasive surgery.[8] Patients who underwent laparoscopic/ robotic kidney transplantation experienced less postoperative pain, less need for analgesics, and a faster recovery. These findings were confirmed in 2021 by Ahlawat et al. who compared the outcomes of RAKT ($n = 126$) versus open KT ($n = 378$). Ischemia time was longer with the robotic approach, however, after a median follow-up period of 2 years, graft outcomes were comparable. Patients who underwent RAKT had less intraoperative blood loss, lower rates of wound complications (0% vs 4%). Overall, these patients experienced a less morbid procedure with less postoperative pain and the need for narcotics.[9]

Obesity may independently be associated with poor graft outcomes and a higher rate of wound complications after open KT.[3–7] However, in patients who

Single-Port Robotic Surgery in Urology. DOI: https://doi.org/10.1016/B978-0-323-91906-7.00010-7

underwent RAKT, similar rates of wound-related complications (2% vs 1.4%) and graft survival were shown in obese versus nonobese recipients.[10]

With the application of single-port (SP) robotic surgery for various urological procedures at many centers, it has been shown that atleast for radical prostatectomy, SP robotic approach would be associated with a shorter hospital stay, less postoperative pain, and quicker recovery compared to conventional multiport robotic approach.[11,12] Therefore the use of a SP robotic platform may be a step forward to perform kidney transplantation even less invasive.

9.2 Single-port robotic renal transplantation—implementation of an innovative procedure

Understandably, SP robotic renal transplantation is an innovative and complex procedure. For the implementation and development of this procedure, the authors recommend following the five-step process for the development of new surgical procedures (IDEAL Model: http://www.ideal-collaboration.net/framework model describes a five-stage process that should be passed to implement a new surgical innovation):

Preclinical (feasibility) step (Phase 0):[13] Practicing and standardizing the procedure in the cadaveric laboratory are highly recommended before the clinical implementation of this innovative approach. This preclinical step is vital in minimizing errors and gaining proficiency in SP RAKT as a new procedure. This preclinical step is also very helpful in assembling and orienting motivated well-versed teams of anesthesiologists, operative room staff, and surgeons skilled in robotic surgery and open kidney transplantation.

Innovation, (Phase 1) (Innovation): In this phase, the initial concept of a new procedure is shown by a few surgeons and applied to very few patients. To our knowledge at present time, SP RAKT is still in this phase.[14]

Development (Phase 2a): More surgeons and innovators demonstrate the safety, feasibility, and reproducibility of the new surgical procedure on more patients.

Exploration (Phase 2b): Diffusion of the procedure to the community by sharing the experiences of many surgeons. This would be helpful in the development of a patient registry and database.

Assessment (Phase 3): Conduction of clinical trials to evaluate the clinical efficacy and cost-effectiveness of the new procedure compared to standard of care.

Long-term studies (Phase 4): For assessment of long-term outcomes and quality assurance.

9.2.1 Patient selection

In our initial experience,[14] we applied SP RAKT in two different scenarios; (1) Adult patients with end-stage renal disease ($n = 6$) and (2) Adult patients who underwent kidney autotransplantation, (ipsilateral SP robotic nephrectomy and kidney transplantation) ($n = 3$).

We excluded patients with complex vascular graft anatomy requiring more than one vascular anastomosis and those requiring a native nephrectomy. Preoperative noncontrast computerized tomography should be used to evaluate the anatomy and configuration of external iliac arteries (with regards to the presence of significant atherosclerotic plaques).

In our initial experience, we used both living and deceased donors. Deceased donors were brain dead donors with a high Kidney Donor Profile Index (> 50%), with minimal (if any) aortic plaque, and without any vascular abnormalities.

As discussed later, both transperitoneal or extraperitoneal approaches can be used to prepare recipients' vasculature and transplant the kidney.

9.2.2 Surgical technique

9.2.2.1 Positioning, access and port placement

The patient is in a supine position with 15 degrees Trendelenburg and 30 degrees rotated to right lateral position. For the transperitoneal approach, the peritoneal cavity is entered directly via a 5 cm midline periumbilical incision (Fig. 9.1). Alternatively, the extraperitoneal approach can be elected and through the same

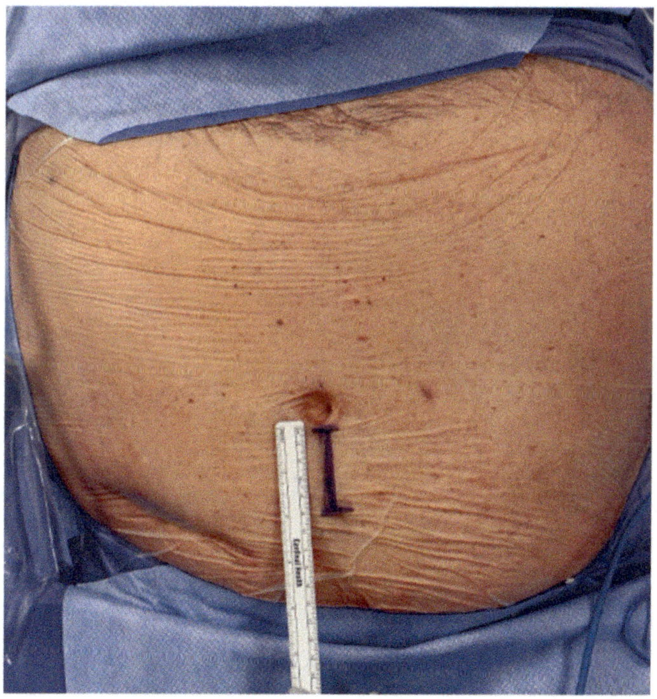

FIGURE 9.1

Intraoperative photograph of the 5 cm midline periumbilical incision.

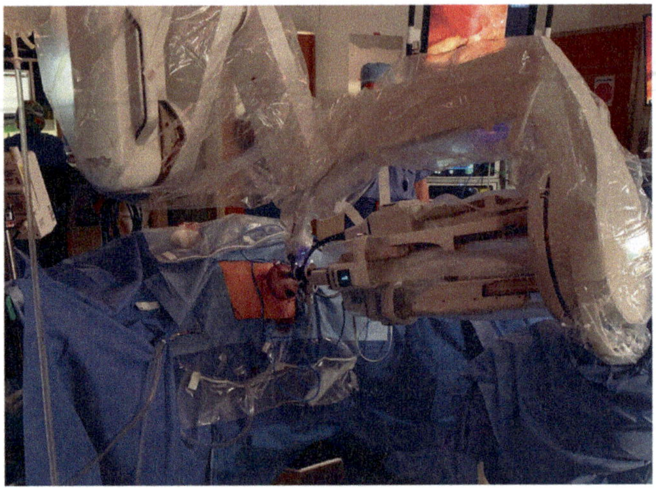

FIGURE 9.2

The single-port robot docked after access port insertion.

incision, the retroperitoneal space is developed by the medial reflection of the peritoneum for direct access to the iliac vessels.

Then a large GelPOINT (Applied Medical, Rancho Santa Margarita, California, United States) system is placed in the wound as the entry point of robotic arms and for wound retraction. The GelPOINT system is also used as the entry point for kidney graft.

The SP trocar and a 12 mm assistant port are inserted through the GelSeal cap, and after establishing the pneumoperitoneum, the SP robot is docked (Fig. 9.2). With the transperitoneal approach, the cecum and right colon will be medialized to expose the external iliac vessels. However, as mentioned above, with an extraperitoneal approach, direct access to the external iliac vessels is possible. The external iliac artery and vein as well as the right anterolateral aspect of the bladder are dissected and prepared for grafting (Fig. 9.3).

9.3 Kidney benching (extracorporeal graft preparation)

After defatting the kidney graft, its pedicle is fully dissected into the hilum and all small vascular branches are tied. Euro-Collins solution is used to flush the renal artery till the efflux is clear. A surgical glove is used as a jacket to wrap the graft in; this would facilitate intracorporeal manipulation with the robotic instruments.[14] To prevent graft torsion during intracorporeal suturing, the upper pole and the posterior surface of the graft are marked with silk stitches on the glove jacket. Similarly, to prevent any vascular torsion, the posterior surface of the renal artery and vein is marked with a marking pen (Fig. 9.4).

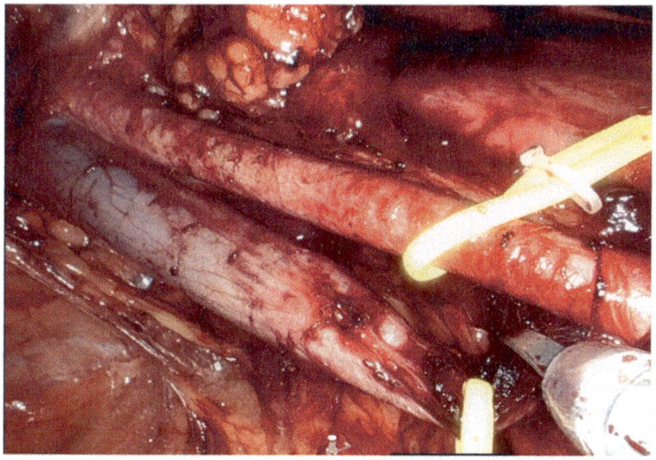

FIGURE 9.3

The external iliac artery and vein are exposed and prepared for grafting.

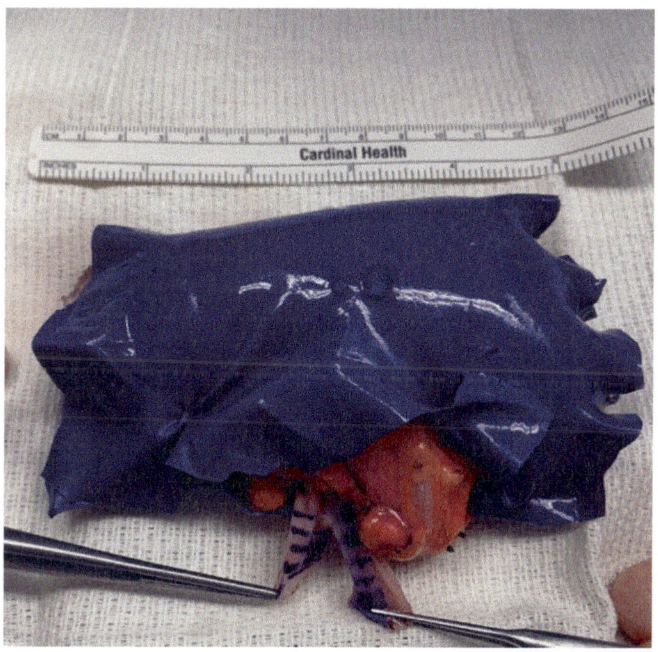

FIGURE 9.4

Photograph of the kidney graft before placing it into the pelvic cavity. Note that the graft is wrapped into the glove jacket and the posterior surface of the artery and vein are marked.

After completion of kidney benching and preparation of iliac vessels, the SP robot is undocked, the GelSeal cap is removed and the graft in its ice jacket is introduced to the pelvic cavity through the port.

9.3.1 Venous anastomosis

With atraumatic robotic bulldog clamps (Scanlan International, Saint Paul, Minnesotta, United States), both proximal and distal parts of the external iliac vein are clamped. A 1 cm venotomy is made on the external iliac vein and the lumen is flushed with heparinized saline solution by a 6 Fr ureteric catheter passed through the assistant port. The renal vein is then anastomosed to the external iliac vein in an end-to-side fashion using GORE-TEX CV-6 suture material (W.L. Gore & Associates Inc. Flagstaff, Arizona, United States) (Fig. 9.5). After completing the anastomosis the renal vein is clamped with a bulldog clamp and the external iliac vein is unclamped cranially and caudally.

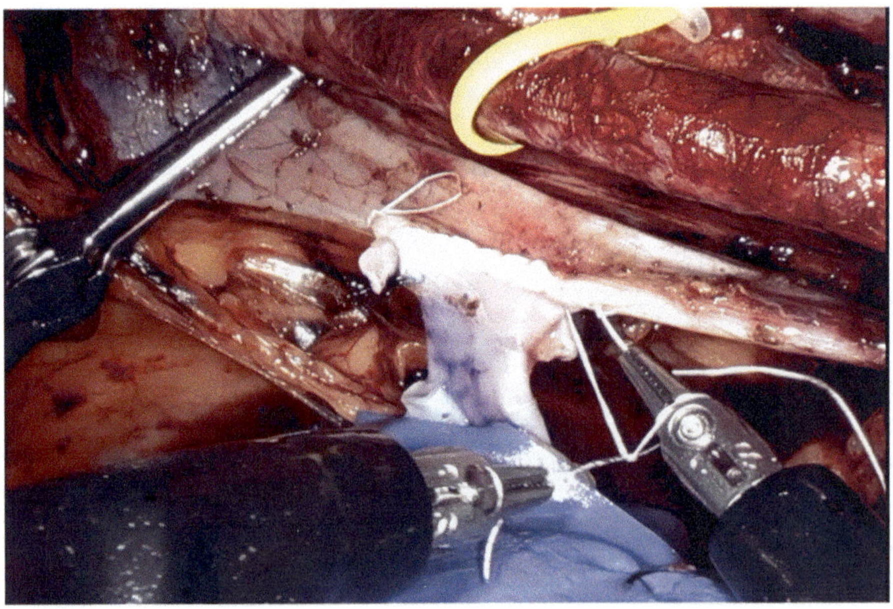

FIGURE 9.5

The graft renal vein is anastomosed to the external iliac vein using GORE-TEX CV-6 suture material.

9.3.2 Arterial anastomosis

After placing bulldog clamps on the proximal and distal portions of the external iliac artery, a 5–7 mm arteriotomy is made by an aortic punch. Alternatively, a protected scalpel can be used to make a linear arteriotomy.

After flushing the iliac artery with heparinized saline, the renal artery can be anastomosed to the external iliac artery in an end-to-side fashion using the same suture material (Fig. 9.6). Similar to the renal vein, the renal artery is clamped with a bulldog before unclamping the external iliac artery. After ensuring hemostasis, the renal vein and artery clamps are removed sequentially. The kidney allograft will be perfused and graft color and turgor should be evaluated.

9.3.3 Ureteroneocystostomy

The kidney is flipped laterally to the prepared retroperitoneal space above the psoas and fixed to the lateral abdominal wall and covered by a peritoneal flap (in case of transperitoneal approach). A 1 cm cystostomy is made on the anterior surface of the bladder. After spatulation of the distal ureter, the ureter is implanted

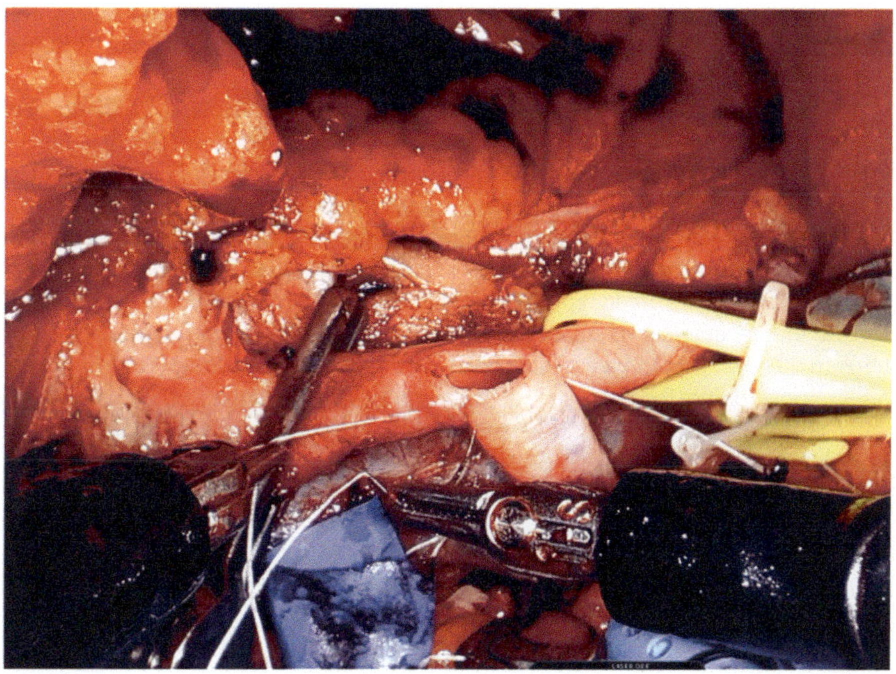

FIGURE 9.6

Anastomosis of the graft renal artery to the external iliac artery is being completed using GORE-TEX CV-6 suture material.

into the bladder by running 4/0 absorbable running sutures over a double J stent (Modified Lich-Gregoir technique). After ensuring the integrity of anastomosis, the robot is undocked and the trocar site is closed in layers.

9.3.4 Technical remarks and outcomes

Menon et al. were the first who standardized the technique of RAKT by application of regional hypothermia in 2014.[15,16] Especially in patients with obesity or in patients with deep and narrow pelvis RAKT would provide a great visualization, exposure, and dexterity.[17,18] European Robotic Urology Section (ERUS) RAKT working group recently reported the largest series of living-donor RAKT ($n = 291$).[19] The mean total surgical time was 244 minutes and the mean rewarming time was 53.1 minutes. The rates of delayed graft function (1.7%), vascular thrombosis (2%), arterial (0.6%), and ureteric (2%) stenosis were very promising.[19]

In general, anastomoses time is longer with robotic compared to open approach, however, this would not negatively affect the long-term graft outcomes. Pein et al. compared the outcome of open ($n = 21$) versus RAKT ($n = 21$) on living allografts: both operative time (+94 minutes) and anastomosis time (+19 minutes) were longer with RAKT. However, graft function and rate of complications were comparable between the two groups after a median follow-up period of 12.9 months.[20]

The technique of SP RAKT is a duplication of *Vattikuti-Medanta* technique[15,16] presented by Menon et al. The SP robotic platform is an adaptable low profile system, well-optimized for multiquadrant surgery. These features are particularly very helpful for renal autotransplantation; both upper tract (Nephrectomy) and pelvic (Transplantation) surgeries as well as graft manipulation can be performed through a SP placed via a small periumbilical incision without the need for undocking. Decaestecker et al. reported the largest series of robotic kidney autotransplantation ($n = 7$) with conventional multiarm robotic systems.[21] They highlighted the need for robot redocking and operative table repositioning for each step of the surgery. Their median operative and vascular anastomosis time were 370 and 28 minutes, respectively without any autograft loss.[21]

In our experience with SP RAKT[14] in patients with ESRD ($n = 6$), 1-month (Cr range = 1.2–1.8 mg/dL, eGFR = 41–55 mL/minute/1.73 m^2) and 3-month (Cr range = 1.5–1.6 mg/dL, eGFR = 45 mL/minute/1.73 m^2) graft functions were comparable with previous reports of conventional RAKT or open surgery.[8,9,17–20]

As shown previously,[11,12] the nonbulky configuration of the SP robotic platform makes it an ideal platform for extraperitoneal robotic surgery. In our initial experience,[14] we were able to transit from the transperitoneal to extraperitoneal approach with this system. The extraperitoneal approach mimics the classic open surgery with direct access to the iliac vessels with minimal (if any) need for manipulation of intraperitoneal organs.

9.4 Limitations, challenges, and future directions

9.4.1 Hypothermia

Unlike open surgery, induction and maintaining intraoperative graft hypothermia during vascular anastomosis has always been a concern during RAKT. Although the use of an ice jacket around the graft is standardized,[15,16] further technical modifications with the use of novel intra-abdominal cooling systems would be interesting.

9.4.2 Vascular anastomosis

Vascular anastomosis is the most critical step during RAKT. According to the ERUS-RAKT working group, the rate of vascular thrombosis after RAKT (2%–2.5%) is comparable to the rate after open surgery (0.5%–3.5%).[18,22,23] To prevent vascular torsion during anastomosis and to avoid disorientation during graft placement, careful graft preparation and marking of the posterior aspects of the allograft and its vasculature are essential.[16,17] Graft and ureter perfusion can be assessed during RAKT by using indocyanine green with the FireFly fluorescence feature.[17]

9.4.3 Logistics

In our initial experience,[14] SP RAKT series were sourced by both deceased and living donors. However, all procedures were done in "elective" circumstances. The technical principles of RAKT are standard independent of the graft source. However, optimizing the recipient, donor, and graft inclusion criteria, as well as logistic coordination, can be challenging in an "emergency" (i.e., deceased donor) situation. The availability of robotic systems and a highly trained dedicated multidisciplinary team are essential for RAKT. Such logistic coordination maybe even more complicated when a SP robotic platform is aimed to be implemented for RAKT from a deceased donor source.

9.4.4 Training

Due to the lack of adequate haptic feedback, during robotic surgery, the force and tension applied on the sutures are visually controlled. As shown, the anastomosis time is generally longer with RAKT compared to open surgery. It has been shown that for experienced surgeons it requires about 35 cases for adequate proficiency especially with regards to an optimized ischemia time with multiarm RAKT.[17] For transition to a SP platform, previous adequate experience with SP robotic surgery would be very valuable. With SP robotic surgery, another learning phase especially with regards to camera control and proficiency in the handling of double-jointed robotic instruments should be considered.

9.4.5 Cost

To our knowledge, there is no cost-analysis study to compare open versus RAKT. A cost-analysis comparative trial between SP versus multiport RAKT would also be very interesting for future investigations. It seems that the operative time and postoperative course may be the main determining factors for variation i n overall cost. Comparative trials with long-term follow-up are required to demonstrate the added value of the SP robotic approach compared to conventional RAKT or open surgery.

9.5 Conclusions

The use of the SP robotic platform is feasible for allotransplantation and autotransplantation. While these early results are promising, larger series with longer follow-up and objective evaluation of outcomes is required. Comparative trials are required to confirm any potential added value of further minimizing the morbidity of kidney transplantation using the SP robotic platform.

References

1. Oberholzer J, Giulianotti P, Danielson KK, et al. Minimally invasive robotic kidney transplantation for obese patients previously denied access to transplantation. *Am J Transplant.* 2013;13(3):721–728.
2. Abaza R, Ghani KR, Sood A, et al. Robotic kidney transplantation with intraoperative regional hypothermia. *BJU Int.* 2014;113(4):679–681.
3. Nanni G, Tondolo V, Citterio F, et al. Comparison of oblique vs hockey-stick surgical incision for kidney transplantation. *Transplant Proc.* 2005;37(6):2479–2481.
4. Humar A, Ramcharan T, Denny R, Gillingham KJ, Payne WD, Matas AJ. Are wound complications after a kidney transplant more common with modern immunosuppression? *Transplantation.* 2001;72(12):1920–1923.
5. Lynch RJ, Ranney DN, Shijie C, Lee DS, Samala N, Englesbe MJ. Obesity, surgical site infection, and outcome following renal transplantation. *Ann Surg.* 2009;250(6):1014–1020.
6. Nashan B, Citterio F. Wound healing complications and the use of mammalian target of rapamycin inhibitors in kidney transplantation: a critical review of the literature. *Transplantation.* 2012;94(6):547–561.
7. Mehrabi A, Fonouni H, Wente M, et al. Wound complications following kidney and liver transplantation. *Clin Transplant.* 2006;17:97–110.
8. Wagenaar S, Nederhoed JH, Hoksbergen AWJ, Bonjer HJ, Wisselink W, van Ramshorst GH. Minimally invasive, laparoscopic, and robotic-assisted techniques vs open techniques for kidney transplant recipients: a systematic review. *Eur Urol.* 2017;72(2):205–217.

9. Ahlawat R, Sood A, Jeong W, et al. Robotic kidney transplantation with regional hypothermia vs open kidney transplantation for patients with end stage renal disease: an ideal stage 2B study. *J Urol*. 2021;205(2):595–602.
10. Prudhomme T, Beauval JB, Lesourd M, et al. Robotic-assisted kidney transplantation in obese recipients compared to non-obese recipients: the European experience. *World J Urol*. 2021;39(4):1287–1298.
11. Kaouk J, Aminsharifi A, Sawczyn G, et al. Single-port robotic urological surgery using purpose-built single-port surgical system: single-institutional experience with the first 100 cases. *Urology*. 2020;140:77–84.
12. Lenfant L, Sawczyn G, Aminsharifi A, et al. Pure single-site robot-assisted radical prostatectomy using single-port vs multiport robotic radical prostatectomy: a single-institution comparative study. *Eur Urol Focus*. 2021;7(5):964–972.
13. Garisto J, Eltemamy M, Bertolo R, Miller E, Wee A, Kaouk J. Single port robot-assisted transperitoneal kidney transplant using the SP® surgical system in a pre-clinical model. *Int Braz J Urol*. 2020;46(4):680–681.
14. Kaouk J, Eltemamy M, Aminsharifi A, et al. Initial experience with single-port robotic-assisted kidney transplantation and autotransplantation. *Eur Urol*. 2021;80 (3):366–373.
15. Menon M, Sood A, Bhandari M, et al. Robotic kidney transplantation with regional hypothermia: a step-by-step description of the Vattikuti Urology Institute-Medanta technique (IDEAL phase 2a). *Eur Urol*. 2014;65(5):991–1000.
16. Menon M, Abaza R, Sood A, et al. Robotic kidney transplantation with regional hypothermia: evolution of a novel procedure utilizing the IDEAL guidelines (IDEAL phase 0 and 1). *Eur Urol*. 2014;65(5):1001–1009.
17. Breda A, Territo A, Gausa L, et al. Robot-assisted kidney transplantation: the European experience. *Eur Urol*. 2018;73(2):273–281.
18. Breda A, Territo A, Gausa L, et al. Robotic kidney transplantation: one year after the beginning. *World J Urol*. 2017;35(10):1507–1515.
19. Musquera M, Peri L, Ajami T, et al. Robot assisted kidney transplantation: update from the European Robotic Urology Section (ERUS) series. *BJU Int*. 2021,127 (2):222–228.
20. Pein U, Girndt M, Markau S, et al. Minimally invasive robotic vs conventional open living donor kidney transplantation. *World J Urol*. 2020;38(3):795–802.
21. Decaestecker K, Van Parys B, Van Besien J, et al. Robot-assisted kidney autotransplantation: a minimally invasive way to salvage kidneys. *Eur Urol Focus*. 2018;4(2): 198–205.
22. Dimitroulis D, Bokos J, Zavos G, et al. Vascular complications in renal transplantation: a single-center experience in 1367 renal transplantations and review of the literature. *Transplant Proc*. 2009;41(5):1609–1614.
23. Vignolini G, Greco I, Sessa F, et al. The University of Florence technique for robot-assisted kidney transplantation: 3-Year experience. *Front Surg*. 2020;7:583798.

CHAPTER

Comparison of perioperative outcomes and costs between single-port and standard multiport robot-assisted surgeries in urology

10

Shirley L. Wang[1], Andrew T. Gabrielson[1], Phillip M. Pierorazio[2] and Mohamad E. Allaf[1]

[1]*The James Buchanan Brady Urological Institute, Johns Hopkins University School of Medicine, Baltimore, MD, United States*

[2]*Division of Urology, Perelman School of Medicine at the University of Pennsylvania, Philadelphia, PA, United States*

10.1 Introduction

Robot-assisted single-port (SP) surgery, which has been increasingly studied in the past decade, represents an opportunity to advance technology and improve patient outcomes. The da Vinci SP (Intuitive Surgical, Sunnyvale, California, United States) is a purpose-built system that received FDA approval in 2018 and follows years of innovation in institution-specific techniques in robotic-laparoendoscopic single-site surgery (R-LESS). The purpose-built SP da Vinci surgical system has enabled broader adoption of this technology among urologic surgeons. Early studies have supported acceptable perioperative outcomes with use of SP robotics. Patient outcomes with a novel surgical technique should be, at minimum, comparable to the standard of care to justify continued use. Here, we review the existing literature on perioperative outcomes and costs associated with SP urologic surgeries with a focus on comparison to MP surgeries.

10.2 Radical prostatectomy

SP-robot-assisted radical prostatectomy (RARP) is the most extensively studied of the SP urologic procedures, with atleast 23 case series and 9 comparative studies, altogether encompassing over 900 patients. Initial studies demonstrate

Single-Port Robotic Surgery in Urology. DOI: https://doi.org/10.1016/B978-0-323-91906-7.00005-3

feasibility and safety of single-site RARP and have since transitioned to assessing differences in patient outcomes based on limited follow-up data (Table 10.1). Although nonstandard approaches to prostatectomy such as perineal and transvesical routes have been explored with SP robotics and shown to be feasible,[1,2] extra- and transperitoneal RARP remain most popular and comprise the bulk of our discussion.

During radical prostatectomies performed by R-LESS or the da Vinci SP system, mean operative times were reported between 114–300 minutes and estimated blood loss between 45 and 350 mL. Length of stay varied between 4.2 hours and 7 days and the rate of Clavien-Dindo complications grade 2 or higher ranged from 0% to 18.4%. Mean lymph node yield ranged from 4–15.9 for cases involving pelvic lymphadenectomy, and rates of positive surgical margin ranged between 0% and 55%. Some authors' postulate high margin rates in the studies were due to high rates of locally invasive disease and utilization of nerve-sparing was pursued where it may not have been appropriate.[3–5] Mean patient-reported measures of pain using the visual analog pain scale (VAPS) were 1.6–2.5 out of 10 in the postoperative period. Erectile function and urinary continence ranged from 7% to 48% and 20% to 100% at 12 weeks after surgery, respectively (Table 10.1).

When comparing mean operative time between MP and SP robotics, results were mixed: two studies showed an increase in mean operative time with SP by 11–15 minutes,[3,4] while others showed decrease with SP by 25–30 minutes.[5,6] At present, the preponderance of evidence suggests that in the hands of an experienced robotic surgeon, use of the SP system does not substantially increase operative time and may even allow diminished operative time, although confounders such as conducting pelvic lymph node (PLND) dissection have not been adequately controlled for. Estimated blood loss was reduced with SP compared to MP surgery in several comparative studies,[4,7,8] and rates of Clavien-Dindo complications grade 2 or higher were similar between both groups. Patients receiving SP-RARP had significantly reduced lengths of stay when compared to MP-RARP.[3,5–7,9] In the largest cohort to date, Lenfant et al. reported same day discharge in 67% of patients with SP versus 7.3% of patients with MP ($P < 0.001$) with median length of stay 4.2 versus 26.1 hours ($P < 0.001$).[7] In an analysis of factors associated with inpatient versus outpatient management of patients with SP-RARP showed two modifiable factors, operative time, and postoperative opioid use, to be significantly associated with same-day discharge.[10] However it is likely that patient selection may have played a role in these findings. Two studies showed decreased mean lymph node yield with the use of SP.[8,9] Other studies including larger cohorts demonstrated no difference in lymph node yield between groups,[6,7,11] suggesting that the SP robotic system does not limit lymph node dissection after surpassing an initial learning curve. Rates of positive surgical margins also did not differ significantly between the two surgical systems in comparative studies (Table 10.1).[4,6–9,11]

With regard to pain, most comparative studies have described similarly low pain scores among SP versus MP groups, with one analysis showing over twice

Table 10.1 Perioperative outcomes of single-port robot-assisted radical prostatectomy.

	Year	Author	Title	n (R-SP RP)	Operative time (min)	EBL (mL)	# of nodes	LoS (days)	Clavien-Dindo complicatio $n \geq 2$ rate	Pain scale (VAPS), OMEQ, or % pain-free at DC	+ Margin rate	Erectile function	Continence rate (0–1 pads at 12 weeks)
R-LESS	2009	Kaouk et al.	Robotic single-port transumbilical surgery in humans: initial report	1	300	250	NR	1.5	0 (0%)	0 at 1 week	0 (0%)	1 (100%) at 3 weeks	0 (0%) at 3 weeks
	2009	Barrett et al.	Re: Initial complete laparoscopic single-site surgery robotic assisted radical prostatectomy (LESS-RARP)	1	210	300	NR	NR	0 (0%)	NR	0 (0%)	1 (100%)	1 (100%)
	2010	White et al.	Robotic laparoendoscopic single-site radical prostatectomy: technique and early outcomes	20	137.5	128.8	4	2.5	3 (15%)	1.6 VAPS, 22.3 OMEQ	4/17 (23%)	NR	9/11 (81%) at 16 weeks
	2012	White et al.	Robotic laparoendoscopic single site urological surgery: analysis of 50 consecutive cases	20	139.5	142	4	2.7	3 (15%)	1.4	4 (20%)	1 SHIM > = 21 at 3 months f/u; 2 at 1 year follow up	18 (90%) at 4 months
	2014	Kaouk et al.	A novel robotic system for single-port urologic surgery: first clinical investigation	11	239	350	4.5	5	3 (27.3%)	NR	2 (18.2%)	1/11 (9%) at 1 year	6/11 (54.5%) at 1 month, 10/11 (91%) at 1 year
	2016	Kaouk et al.	Descriptive technique and initial results for robotic radical perineal prostatectomy.	4	230	62.5	NR	1.4	0 (0%)	NR	3 (75%)	NR	4 (100%) at 6 months

(Continued)

Table 10.1 Perioperative outcomes of single-port robot-assisted radical prostatectomy. *Continued*

Year	Author	Title	n (R-SP RP)	Operative time (min)	EBL (mL)	# of nodes	LoS (days)	Clavien-Dindo complicatio $n \geq 2$ rate	Pain scale (VAPS), OMEQ, or % pain-free at DC	+ Margin rate	Erectile function	Continence rate (0–1 pads at 12 weeks)
2017	Tugcu et al.	Single plus one port robotic radical prostatectomy (SPORP); Initial experience.	8	143	45	NR	NR	NR	NR	0 (0%)	NR	6 (75%)
2018	Mattevi et al.	First case of robotic laparoendoscopic single-site radical prostatectomy with single-site VesPa platform.	1	300	400	NR	NR	0 (0%)	NR	0 (0%)	0 (0%)	0 (0%) at 12 weeks; 1 (100%) at 1 year
2018	Tugcu et al.	Robotic perineal radical prostatectomy and robotic pelvic lymph node dissection via a perineal approach: the Tugcu Bakirkoy technique.	7	184.1	64.2	NR	2.1	NR	NR	1 (14%)	NR	NR
2019	Gaboardi et al.	Robotic laparoendoscopic single-site radical prostatectomy (R-LESS-RP) with daVinci Single-Site (R) platform. Concept and evolution of the technique following an IDEAL phase 1.	12	256.2	165	NR	6.54	5 (41.7%)	NR	1 (8.3)	4/5 (80%) IIEF> 17 at mean 1 year f/u	12 (100%) at mean 1 year f/u

Case series	2019	Chang et al.	Single-port transperitoneal robotic-assisted laparoscopic radical prostatectomy (spRALP): initial experience	1	152	100	NR	4	NR	NR	0 (0%)	NR	NR
	2019	Kaouk et al.	Single-port robot-assisted radical prostatectomy: first clinical experience using the SP surgical system	2	140	NR	NR	NR	NR	NR	NR	NR	NR
	2019	Kaouk et al.	Extraperitoneal single-port robot-assisted radical prostatectomy: initial experience and description of technique	10	197.5	143	10.1	0	0 (0%)	NR	5 (50%)	3 (30%) at 12 weeks	2 (20%) at 12 weeks; 5 (50%) at 90 days
	2019	Dobbs et al.	Single-port robot-assisted laparoscopic radical prostatectomy: initial experience and technique with the da Vinci SP platform	10	234	65	NR	1	0 (0%)	NR	2 (20%)	NR	NR
	2019	Ng et al.	Robot-assisted single-port radical prostatectomy: a phase 1 clinical study	20	209	296	8.3	5	3 (15%)	1.9 at 2 weeks	11 (55%)	NR	NR

(*Continued*)

Table 10.1 Perioperative outcomes of single-port robot-assisted radical prostatectomy. *Continued*

Year	Author	Title	n (R-SP RP)	Operative time (min)	EBL (mL)	# of nodes	LoS (days)	Clavien-Dindo complicatio $n \geq 2$ rate	Pain scale (VAPS), OMEQ, or % pain-free at DC	+ Margin rate	Erectile function	Continence rate (0–1 pads at 12 weeks)
2019	Steinberg et al.	Magnet-assisted robotic prostatectomy using the da Vinci SP robot: an initial case series	15	224	198	10.25	1.25	1 (6.7%)	NR	2 (13%)	NR	NR
2020	Kaouk et al.	Robotic urologic surgical interventions performed with the single port dedicated platform: first clinical investigation.	3	227	83	15	0.67	0 (0%)	NR	NR	NR	NR
2020	Agarwal et al.	Initial experience with da Vinci single-port robot-assisted radical prostatectomies	49	162	191	7.75	1.25	4 (8.2%)	NR	13 (26.5%)	10 (48%) at 12 weeks	16/21 (76%)
2020	Kim et al.	Single-port robot-assisted radical prostatectomy with the da Vinci SP system: a single surgeon's experience	20	245	200	19	7	0 (0%)	NR	7 (35%)	4/7 (57%)	10/10 (100%)
2020	Kaouk et al.	Single-port robotic urological surgery using purpose-built single-port surgical system: single-institutional experience with the first 100 cases	60	104–281	50–200	5	0.14–2.26	6 (10%)	41 (68.3%) opioid-free pain management	18 (30%)	NR	51.1% at 30 days

	2020	Wilson et al.	Outpatient extraperitoneal single-port robotic radical prostatectomy	60	198	179	NR	0.18	7 (11.7%) ≥ 3	VAPS 2, 37% pain free at DC	23%	1/19 (5.26%) SHIM >19 at 90 days	37 (76%) at 90 days
	2020	Aminsharifi et al.	Predictors associated with a prolonged hospital stay after single-port extraperitoneal robotic radical prostatectomy: a comparative analysis of outpatient versus inpatient care	120	203.2	197.2	NR	0.19	7 (5.8%)	VAPS 2 at discharge	29 (24.2%)	NR	102 (85.0%) at 90 days
	2020	Covas Moschovas et al.	Early outcomes of single-port robot-assisted radical prostatectomy: lessons learned from the learning-curve experience	50	118	50	NR	0.68	0 (0%)	2 VAPS 8 h, 0 at 16 h	7 (14%)	70% at 6–8 weeks	27 (54%) at 4 weeks
	2021	Kaouk et al.	Single-port transvesical robotic radical prostatectomy: initial										
Clinical experience and description of technique	20	199	135	NR	0.18	1 (5%)	VAPs 2.5, 0 opioid use at DC	3 (15%)	NR	15 (75%) at Foley removal; 17 (85%) at 10 days			

(Continued)

Table 10.1 Perioperative outcomes of single-port robot-assisted radical prostatectomy. *Continued*

	Year	Author	Title	n (R-SP RP)	Operative time (min)	EBL (mL)	# of nodes	LoS (days)	Clavien-Dindo complicatio $n \geq 2$ rate	Pain scale (VAPS), OMEQ, or % pain-free at DC	+ Margin rate	Erectile function	Continence rate (0–1 pads at 12 weeks)
Comparative	2020	Kaouk et al.	Extraperitoneal versus transperitoneal single port robotic radical prostatectomy: a comparative analysis of perioperative outcomes	52 (SP)	201*	145.7	5*	0.18	6 (11.5%)	NR	14 (26.9%)	NR	24 (60%)
				46 (MP)	248.2*	117.6	12*	1.07**	7 (15.2%)	NR	19 (41.3%)	NR	25 (62.5%)
	2020	Saidian et al.	Perioperative outcomes of single versus multi-port robotic assisted radical prostatectomy: a single institutional experience	47 (SP)	255.9	169.2	15.9	1.1	2 (4.3%)	NR	10 (21.3%)	3 (7.9%) at 12 weeks	19 (63.3%)
				48 (MP)	274.7	157.7	15.1	1.4	3 (6.3%)	NR	13 (27.1)	3 (6.7%) at 12 weeks	24/45 (53.3%)
	2020	Huang et al.	Comparison of perioperative and pathologic outcomes between single-port and standard robot-assisted radical prostatectomy: an analysis of a high-volume center and the pooled world experience	26 (SP)	180	100*	5*	1	3 (12%)	VAPS 2	6 (23.1%)	NR	NR
				376 (MP)	180	150*, $P = 0.04$	9*, $P = 0.002$	1	47 (13%)	VAPS 3	92 (24%)	NR	NR

2020	Vigneswaran et al.	A comparison of perioperative outcomes between single-port and multiport robot-assisted laparoscopic prostatectomy	50 (SP)	260	100	12*	1*	7 (14%)	29 OMEQ, 15 (30%) pain-free POD1*	20/48 (41.7%)	NR	24 (86%)
			113 (MP)	280	100	14*, $P = 0.012$	2*, $P = 0.002$	17 (15%)	28 OMEQ, 13 (12%) pain-free POD1, $P = 0.004$	47/113 (41.6%)	NR	84 (87%)
2020	Lenfant et al.	Single-institution cost comparison: single-port versus multiport robotic prostatectomy	78 (SP)	198*	NR	5.18	0.41*	NR	NR	20 (26%)	NR	NR
			97 (MP)	223*, $P < 0.001$	NR	5.86	1.48*, $P < 0.001$	NR	NR	38 (39%)	NR	NR
2021	Francavilla et al.	Radical prostatectomy technique in the robotic evolution: from da Vinci standard to single port—a single surgeon pathway	40 (SP)	238.5*	75*	12	1*	5 (12.5%) ≥ 3	NR	15 (38.5%)	NR	25/37 (67.6%) at 90 days
			40 (MP)	270.5*, $P < 0.01$	88*, $P < 0.01$	9	2*, $P < 0.01$	5 (12.5%) ≥ 3	NR	21 (52.5%)	NR	27/36 (75.0%) at 90 days
2021	Abaza et al.	Single-port robotic surgery allows same-day discharge in majority of cases	59 (SP)	170*	Not reported for each procedure.							
			1131 (MP)	155.5*, $P < 0.001$	Not reported for each procedure.							

(*Continued*)

Table 10.1 Perioperative outcomes of single-port robot-assisted radical prostatectomy. *Continued*

	Year	Author	Title	n (R-SP RP)	Operative time (min)	EBL (mL)	# of nodes	LoS (days)	Clavien-Dindo complicatio $n \geq 2$ rate	Pain scale (VAPS), OMEQ, or % pain-free at DC	+ Margin rate	Erectile function	Continence rate (0–1 pads at 12 weeks)
	2019	Lenfant et al.	Pure single-site robot-assisted radical prostatectomy using single-port versus multiport robotic radical prostatectomy: a single-institution comparative study	100 (SP)	195	190*	5	0.18*	12 (12%)	7.5 OMEQ*	24 (24%)	20%	60 (69%) at 12 weeks; 34 (85%) at 1 year
				110 (MP)	190	200*, $P < 0.0001$	6	1.09*, $P < 0.0001$	6 (5.5%)	15 OMEQ*, $P < 0.0001$	37%	25%	60 (73%) at 12 weeks; 36 (88%) at 1 year
	2021	Covas Moschovas et al.	Comparing the approach to radical prostatectomy using the multiport da Vinci Xi and da Vinci SP robots: a propensity score analysis of perioperative outcomes	71 (SP)	114*	55 (78%) under 100 mL*	NR	NR	0 (0%)	VAPS 2.5 at 18 h	12 (17%)	18% at 90 days	68% at 90 days
				71 (MP)	103, $P < 0.001$	36 (51%)*, $P = 0.002$	NR	NR	0 (0%)	2	11 (16%)	28%	56%

Statistically significant difference 2

as many SP patients being pain-free on postoperative day 1 (30% vs 12%, $P = 0.004$).[9] In the largest cohort of patients undergoing SP-RARP, Lenfant et al. demonstrated substantially decreased rates of opioid use both during admission and following discharge among SP patients (32% vs 64%, $P < 0.001$ and 35% vs 88%, respectively).[7]

Functional outcomes are also an important consideration when evaluating a new surgical technique. Few comparative studies have evaluated erectile function and urinary continence following SP-RARP, with no significant differences with upto 12 weeks of follow-up.[4,7,11] Nerve-sparing prostatectomy is feasible with the SP robotic system, with comparable rates performed between the two groups.[9]

Since several years, the feasibility and safety of SP-RARP has been proven, and more recently, the literature has shifted towards evaluation of this system in comparison to the current standard of care, MP-RARP. Recent comparative studies support non-inferior peri-operative outcomes and in some cases, shortened length of stay and reduced required pain management. As the field moves forward, randomized trials should be conducted to critically evaluate differences in perioperative outcomes observed in studies thus far.

10.3 Radical and partial nephrectomy

Since the advent of robotic single-site surgery in 2008, more than 200 partial and radical nephrectomies have been reported using SP robotics.[12,13] Most of these cases have been published in small case series which report a variety of perioperative variables, the most salient of which will be highlighted here (Table 10.2). The bulk of SP robotics experience in renal surgery involves use of this technology in partial nephrectomy. Importantly, these series involve selected patients with low-complexity clinical stage T1a renal masses, and thus are not readily comparable to studies measuring perioperative outcomes with hand-assisted or MP robotic renal surgery which include a heterogenous mix of tumors. Mean operative times for SP robotic renal surgery ranged from 105–233 minutes and mean warm ischemia time ranged from 26–38 minutes for those undergoing partial nephrectomy. Mean hospital stay was 1.7–5.2 days. Complication rates were generally low with most series reporting 0%–12.5% Clavien-Dindo complications of 2 or higher. Mean estimated blood loss ranged from 68–464 mL. Several studies also included a pain metric in the form of VAPS at 1 week postoperative, ranging from 1–2 out of 10. One study demonstrated that compared to MP robotics, cases performed with the SP system had significantly reduced opiate medication requirements postoperatively, however the study was limited by small sample size.[14] Positive margin rate was low and consistent with results reported in MP robotic renal surgery (Table 10.2).

Several retrospective studies provided comparisons of perioperative outcomes for patients undergoing SP robotic renal surgery versus conventional MP robotic

Table 10.2 Perioperative outcomes of single-port robot-assisted nephrectomy.

	Year	Author	Title	n (R-SP renal surgery)	Operative time (min)	EBL (mL)	Warm ischemia time (min)	LoS (days)	Clavien-Dindo complication ≥ 2 rate	Pain (VAPS)	+ Margin rate
Case series	2009	Kaouk et al.	Robotic SP transumbilical surgery in humans: initial report	1	150	200	NR	2	0 (0%)	0	0 (0%)
	2009	Kaouk et al.	SP laparoscopic and robotic PN	2	170	100	NR	3.5	0 (0%)	1	0 (0%)
	2013	Stein et al.	R-LESS using GelPort as the access platform	2	190	425	NR	1.75	0 (0%)	NR	NR
	2011	Kaouk et al.	LESS in urology: worldwide multiinstitutional analysis of 1076 cases	143	NR specific to SP						
	2011	Han et al.	Robot-assisted laparoscopic single-site surgery: PN for malignancy	16	233	464	30	5.2	2 (12.5%) conversion to open	NR	0 (0%)
	2012	White et al.	R-LESS urological survey: analysis of 50 consecutive cases	24	NR specific to SP renal surgery						
	2013	Tiu et al.	Feasibility of R-LESS PN for renal tumors >4 cm	67	178 for <4 cm, 197 for >4 cm	271 for <4 cm, 408 for >4 cm	24 for <4 cm, 31 for >4 cm	4 for <4 cm, 5.3 for >4 cm	3 (15%) for <4 cm, 7 (14%) for >4 cm	NR	1 (5%) for <4 cm, 2 (4.2) for >4 cm
	2014	Mathieu et al.	Robotic-assisted LESS-RN: first experience with the novel da Vinci single-site platform	6	179	100	N/A	3	0 (0%)	NR	0 (0%)
	2014	Komninos et al.	Robotic-assisted LESS PN with the novel da Vinci single-site platform: initial experience	3	105	250	NR	4.5	0 (0%)	NR	0 (0%)
	2014	Kaouk et al.	A novel robotic system for SP urologic surgery: first clinical investigation	8	157–232	100–550	38	5	1 (12.5%)	NR	0 (0%)
	2020	Fang et al.	Single-port robotic partial and radical nephrectomies for renal cortical tumors: initial clinical experience	16	176	50 for RN, 200 for PN	All off-clamp	2 for RN, 3 for PN	2 (67%) for RN, 3 (23%) for PN	NR	1 (11%)
	2021	Shukla et al.	Single-port robotic-assisted partial nephrectomy: initial clinical experience and lessons learned for successful outcomes	12	172	68	25	1.2	0 (0%)	NR	1 (8.3%)

	2021	Francavilla et al.	Single-port robot assisted partial nephrectomy: initial experience and technique with the da Vinci Single-Port platform (IDEAL Phase 1)	14	202	NR	20	1	2 (14%)	NR	1 (7%)
Comparative	2011	White et al.	R-LESS RN: surgical technique and comparative outcomes	10 (SP)	167.5	100	N/A	2.5	1 (10%)	2	0 (0%)
				10 (Conventional lap)	150	100	N/A	3	1 (10%)	0	0 (0%)
	2014	Komninos et al.	R-LESS PN trifecta outcome is inferior to multiport robotic PN: comparative analysis	78 (SP)	208	335	26.5	4.6	7 (4%)	NR	1 (1.8%)
				80 (MP)	173	313	20.2	5	4 (2.3%)	NR	4 (4.8%)
	2014	Shin et al.	LESS FAPN reduces postoperative wound pain without rise in complication rates (same cohort as above)	78 (SP)	NR	NR	NR	NR	NR	1.7	NR
				80 (MP)	NR	NR	NR	NR	NR	2.1	NR

surgery.[12] The preponderance of evidence demonstrated no significant difference between SP and MP robotic renal surgery with respect to estimated blood loss, positive surgical margin rate, postoperative renal function, and perioperative complication rate.[12] Use of SP robotics has been associated with lower analgesic requirements and decreased postoperative pain based on VAPS, as well as improved cosmesis and shorter length of stay in hospital.[14] There have been mixed results regarding whether total operative and warm ischemia time are significantly different with SP robotic renal surgery. Whether increases in operative and warm ischemia time is secondary to intrinsic limitations of the SP system or reflects the steep learning curve involved with transitioning from MP to SP systems has yet to be adequately assessed. When taken together, SP robotics appears to offer improved convalescence and noninferior oncologic outcomes when compared to MP robotics, however further studies are needed to better define the utility of SP systems in partial and radical nephrectomy (Table 10.2).

10.4 Radical cystectomy

To date, five case series have described SP radical cystectomy with intracorporeal urinary diversion including a total of 26 patients (Table 10.3). Reported operative times ranged from 330–556 minutes, with estimated blood loss between 100 and 750 mL. The length of stay for these patients was 4–5.5 days. Rates Clavien-Dindo complications of grade 2 or greater ranged from 17% to 41.7%. In their most recent series of six patients, Kaouk et al. report a Clavien-Dindo grade IIIa complication (lymphocele) in one of the six patients.[15] The mean lymph node yield ranged from 8.5 to 12.5. Positive margin rate is reported in only two studies: one case (8.3%) in the series by Gross et al.,[16] and none in the series by Kaouk et al.[15] Additionally, recurrence is reported in only one study, with one patient of six in the series by Kaouk et al. found to have distant metastasis of 6 months following surgery.[15]

Gross et al. are the first group to perform a comparative study of SP RARC versus MP RARC.[16] This retrospective analysis included 12 patients who underwent a procedure with the SP system, each of whom was propensity matched to two patients who underwent a procedure with the MP system. Those receiving an SP RARC had a lower mean lymph node yield (11.9 vs 17.1, $P = 0.035$). The authors attribute this to a learning curve associated with SP PLND dissection, similar to that of MP PLND. All other measured perioperative outcomes, including operative time, EBL, LOS, positive surgical margin, and 90-day complication rate were similar between the two groups (Table 10.3).

Overall, existing case series support feasibility and safety in the immediate perioperative period of SP RARC. Further study is required to assess feasibility among a wider group of surgeons and oncologic safety, particularly given decreased lymph node yield observed in the sole comparative study of SP RARC.

Table 10.3 Perioperative outcomes of single-port robot-assisted radical cystectomy.

Year	Author	Title	n (R-SP RC)	Operative Time (min)	Estimated Blood Loss (mL)	# of nodes	LoS (days)	Clavien-Dindo Complication ≥2 rate	Pain Scale (VAPS)	Positive margin rate	Recurrence rate
2012	White et al.	Robotic laparoendoscopic single site urological surgery analysis of 50 consecutive cases	3	290	266.7	16	7.7	1 (33%)	1.3 VAPS	0 (0%)	NR
2019	Kaouk et al.	Single-port robotic intracorporeal ileal conduit urinary diversion during radical cystectomy using the SP surgical system: step-by-step technique	3	447	NR	NR	5	0 (0%)	NR	NR	NR
2019	Kaouk et al.	A novel robotic system for SP urologic surgery: first clinical investigation	1	330	100	NR	4	0 (0%)	NR	NR	NR
2020	Zhang et al.	Single port robotic radical cystectomy with intracorporeal urinary diversion: a case series and review	4	270	250	12.5	5.5	0 (0%)	NR	NR	NR

(Continued)

Table 10.3 Perioperative outcomes of single-port robot-assisted radical cystectomy. *Continued*

Year	Author	Title	n (R-SP RC)	Operative Time (min)	Estimated Blood Loss (mL)	# of nodes	LoS (days)	Clavien-Dindo Complication ≥2 rate	Pain Scale (VAPS)	Positive margin rate	Recurrence rate
2020	Kaouk et al.	Single-port robotic urological surgery using purpose-built single-port surgical system: single-institutional experience with the first 100 cases	6	334–556	100–750	8.5	4–5 days	1 (17%)	NR	0 (0%)	1 (17%)
2020	Gross et al.	Initial experience with single-port robot-assisted radical cystectomy: comparison of perioperative outcomes between single-port and conventional multiport approaches	12	387	117	11.9*	5.4	5 (41.7%)	NR	1 (8.3%)	NR
			24	418	148	17.1*, $P = 0.034$	5.9	10 (41.7%)	NR	1 (4.2)	NR

*$p < 0.05$.

10.5 Pyeloplasty

Although SP robotics is being increasingly adopted for radical prostatectomy, radical cystectomy, and radical and partial nephrectomy, it has failed to gain significant traction as a modality in performing pyeloplasty. To our knowledge, only seven studies involving roughly 60 patients have described their experience with SP pyeloplasty (Table 10.4). These initial studies demonstrate that SP robotics can successfully be employed for pyeloplasty with acceptable mean operative times (152–270 minutes), estimated blood loss (10–80 mL), length of hospital stay (1–4 days) and rates of complications (0%–13.3% of Clavien-Dindo classification 2 +). Robotic pyeloplasty was found to shorten the length of hospital stay with noninferior complication and success outcomes in a large systematic review and meta-analysis. SP may be able to replicate these findings but also provide an improved cosmetic outcome which may be especially important to children who undergo this procedure.

10.6 Sacrocolpopexy

Several groups have described their initial experience with and technical considerations for performing robotic-assisted sacrocolpopexy using SP robotics.[17–19] To-date, three case series have been reported involving 34 patients (Table 10.5). SP robotic-assisted sacrocolpopexy appears to be feasible with low complication rates, minimal blood loss and postsurgical pain, short convalescence, and improved cosmesis. One study conducted by Matanes et al., demonstrated significant improvement in the learning curve between the first 15 cases and the following 10 cases.[18] This corresponded to a 1-hour reduction in total operative time. The operative times for the last 10 cases in the series were comparable to those reported for the initial experience with MP robotics.[18]

10.7 Vaginoplasty

More recently, SP robotics has demonstrated safety and feasibility for use in gender-affirming procedures, namely in robotic-assisted peritoneal flap vaginoplasty using a modified Davydov technique.[20] Only one large case series comparing SP and MP robotic peritoneal flap vaginoplasty has been published by Dy and Zhao et al. which included 90 patients utilizing the SP system.[20] Mean operative time was 220 minutes, which was significantly shorter than the procedure performed with the MP system. The authors postulated that this shorter operative time with the SP system was due to improved visualization and mobility of the arms intracorporeally, as well as allowing for more space for another surgeon to perform perineal work concurrently. Mean length of stay with the SP system was

Table 10.4 Perioperative outcomes of single-port robot-assisted pyeloplasty.

Year	Author	Title	n (R-SP pyeloplasty)	Operative time (min)	Estimated blood loss (mL)	LoS (days)	Clavien-Dindo complication>2 rate	Pain factor (VAPS or OMEQ)
2009	Kaouk et al.	Robotic SP transumbilical surgery in humans: initial report	1	270	80	2	0 (0%)	NR
2012	Olweny et al.	Robot-assisted LESS pyeloplasty: technique using the da Vinci Si robotic platform	10	226	56	2.6	1 (10%)	Mean OMEQ 30.3
2013	Tobias et al.	Robot-assisted transumbilical LESS pyeloplasty: technique and perioperative outcomes from a single institution	8	173	25	1	1 (12.5%)	NR
2015	Stein et al.	R-LESS using GelPort as the access platform	2	235	38	1.5	0 (0%)	NR
2015	Buffi et al.	Robotic-assisted, single-site, dismembered pyeloplasty for ureteropelvic junction obstruction with the new da Vinci platform: A Stage 2a study	30	160	NR	5	4 (13.3%)	VAPS 1/10
2019	Kang et al.	Robot-assisted laparoscopic SP pyeloplasty using the da Vinci SP system: initial experience with a pediatric patient	1	211	10	4	0 (0%)	NR
2021	Caddedu et al.	Single-port robotic surgery in urology	12	152	15	2	0 (0%)	NR

Table 10.5 Perioperative outcomes of single-port robot-assisted sacrocolpopexy.

Year	Author	Title	n (R-SP Sacrocolpopexy)	Operative time (min)	Estimated blood loss (mL)	Length of stay (days)	Clavien-Dindo complication >2 rate	Pain factor (VAPS)
2016	Lee et al.	Robotic single-site sacrocolpopexy: first report and technique using the single-site wristed needle driver	6	122	66	3.5	0 (0%)	NR
2017	Matanes et a.	SP robotic-assisted sacrocolpopexy: our experience with the first 25 cases	25	190	30	2	1 (4%)	1
2020	Ganesan et a.	SP robotic-assisted laparoscopic sacrocolpopexy with magnetic retraction: first experience using the SP da Vinci platform	3	225	30	1	0 (0%)	NR

5 days, which was similar to that of the MP robotic approach. Importantly, there was no difference in vaginal depth between the two approaches. SP was associated with reduced need for perioperative transfusions and lower rates of vaginal stenosis; however, this could be attributed to shorter follow-up in the SP arm as well as the surgeon learning curve occurring using the MP rather than SP system.

10.8 Cost analysis of single-port robotics

Perioperative outcomes of the more thoroughly studied SP urologic procedures have been demonstrated to be largely noninferior to their MP counterparts. In addition to equivalent or improved outcomes, innovative surgical techniques should remain cost-effective for both the patient and the healthcare system.

To date, only one study has evaluated differences in costs associated with SP- and MP-RARP.[5] In a single-institution retrospective analysis, Lenfant et al. found mean total cost between the SP- and MP-RARP did not differ significantly ($13,512 ± $1,615 vs $13,284 ± $1360, $P = 0.32$). SP was associated with increased cost of disposables in the operating room ($2445 ± $1096 vs $763 ± $310, $P < 0.001$); however, these differences were offset by decreased cost of hospitalization due to significantly shorter length of stay in the SP group (9.84 ± 11.3 hours vs 35.5 ± 29.1 hours, $P < 0.001$).

This early study demonstrates costs of SP- and MP-RARP are comparable in a high-volume setting with an experienced robotic surgeon. Currently, there are no studies evaluating cost comparison of other SP procedures. Extrapolation of these findings suggest that, with similar perioperative outcomes and shortened hospitalizations, other SP procedures may be economically viable. Further evaluation should include longer-term cost comparison between SP- and MP-RARP, applicability to other procedures, as well as assessment of costs to the patient.

10.9 Conclusion

The da Vinci SP platform represents the newest evolution in technology used in robotic-assisted surgery and has demonstrated viability for several major urological operations. The current, albeit limited literature describing the use of the SP system in urology has shown equivalent perioperative, oncologic, functional outcomes following surgery when compared to MP systems. Although preliminary comparisons of cost-effectiveness, cosmesis, and convalescence have demonstrated noninferior outcomes between SP and MP systems, these comparisons require further analysis in prospective cohorts. It is important to note that this chapter focuses strictly on published series and is thus limited by the paucity of studies describing outcomes following SP urologic surgery. However, as SP technology becomes more accessible to urologists, so too will the field have access to more mature data that can drive meaningful comparisons between SP and MP robotics. Even in its infancy, SP robotics are being increasingly utilized for complex oncologic and reconstructive surgeries. As urologists continue to push for less morbid and more minimally invasive surgical techniques, we anticipate SP surgery to play a growing role in the delivery of urologic surgical care.

Conflicts of interest and financial disclosures

None

References

1. Kaouk JH, Akca O, Zargar H, et al. Descriptive technique and initial results for robotic radical perineal prostatectomy. *Urology*. 2016;94:129–138. Available from: https://doi.org/10.1016/j.urology.2016.02.063.
2. Kaouk J, Beksac AT, Zeinab MA, Duncan A, Schwen ZR, Eltemamy M. Single port transvesical robotic radical prostatectomy: initial clinical experience and description of technique. *Urology*. May 2021. Available from: https://doi.org/10.1016/j.urology.2021.05.022. S0090429521004301.
3. Abaza R, Murphy C, Bsatee A, Brown DH, Martinez O. Single-port robotic surgery allows same-day discharge in majority of cases. *Urology*. 2021;148:159–165. Available from: https://doi.org/10.1016/j.urology.2020.08.092.
4. Moschovas MC, Bhat S, Sandri M, et al. Comparing the approach to radical prostatectomy using the multiport da Vinci Xi and da Vinci SP robots: a propensity score analysis of perioperative outcomes. *Eur Urol*. 2021;79(3):393–404. Available from: https://doi.org/10.1016/j.eururo.2020.11.042.
5. Lenfant L, Sawczyn G, Kim S, Aminsharifi A, Kaouk J. Single-institution cost comparison: single-port vs multiport robotic prostatectomy. *Eur Urol Focus*. 2020. Available from: https://doi.org/10.1016/j.euf.2020.06.010. Published online July 4.
6. Francavilla S, Veccia A, Dobbs RW, et al. Radical prostatectomy technique in the robotic evolution: from da Vinci standard to single port-a single surgeon pathway. *J Robot Surg*. 2021. Available from: https://doi.org/10.1007/s11701-021-01194-8. Published online February 7.
7. Lenfant L, Sawczyn G, Aminsharifi A, et al. Pure single-site robot-assisted radical prostatectomy using single-port vs multiport robotic radical prostatectomy: a single-institution comparative study. *Eur Urol Focus*. 2020. Available from: https://doi.org/10.1016/j.euf.2020.10.006. Published online November 4.
8. Huang MM, Patel HD, Wainger JJ, et al. Comparison of perioperative and pathologic outcomes between single-port and standard robot-assisted radical prostatectomy: an analysis of a high-volume center and the pooled world experience. *Urology*. 2021;147:223–229. Available from: https://doi.org/10.1016/j.urology.2020.08.046.
9. Vigneswaran HT, Schwarzman LS, Francavilla S, Abern MR, Crivellaro S. A comparison of perioperative outcomes between single-port and multiport robot-assisted laparoscopic prostatectomy. *Eur Urol*. 2020;77(6):671–674. Available from: https://doi.org/10.1016/j.eururo.2020.03.031.

10. Aminsharifi A, Wilson CA, Sawczyn G, Kim S, Lenfant L, Kaouk J. Predictors associated with a prolonged hospital stay after single-port extraperitoneal robotic radical prostatectomy: a comparative analysis of outpatient vs inpatient care. *J Endourol.* 2020;34(10):1049–1054. Available from: https://doi.org/10.1089/end.2020.0389.
11. Saidian A, Fang AM, Hakim O, Magi-Galluzzi C, Nix JW, Rais-Bahrami S. Perioperative outcomes of single vs multi-port robotic assisted radical prostatectomy: a single institutional experience. *J Urol.* 2020;204(3):490–495. Available from: https://doi.org/10.1097/JU.0000000000000811.
12. Komninos C, Shin TY, Tuliao P, et al. R-LESS partial nephrectomy trifecta outcome is inferior to multiport robotic partial nephrectomy: comparative analysis. *Eur Urol.* 2014;66(3):512–517. Available from: https://doi.org/10.1016/j.eururo.2013.10.058.
13. Tiu A, Kim KH, Shin TY, Han WK, Han SW, Rha KH. Feasibility of robotic laparoendoscopic single-site partial nephrectomy for renal tumors >4 cm. *Eur Urol.* 2013;63(5):941–946. Available from: https://doi.org/10.1016/j.eururo.2012.11.031.
14. White MA, Autorino R, Spana G, et al. Robotic laparoendoscopic single-site radical nephrectomy: surgical technique and comparative outcomes. *Eur Urol.* 2011;59(5):815–822. Available from: https://doi.org/10.1016/j.eururo.2011.02.020.
15. Kaouk J, Aminsharifi A, Sawczyn G, et al. Single-port robotic urological surgery using purpose-built single-port surgical system: single-institutional experience with the first 100 cases. *Urology.* 2020;140:77–84. Available from: https://doi.org/10.1016/j.urology.2019.11.086.
16. Gross JT, Vetter JM, Sands KG, et al. Initial experience with single-port robot-assisted radical cystectomy: comparison of perioperative outcomes between single-port and conventional multiport approaches. *J Endourol.* 2021;35(8):1177–1183. Available from: https://doi.org/10.1089/end.2020.1227. Published online April 5.
17. Lee SR. Robotic Single-Site® sacrocolpopexy: first report and technique using the Single-Site® wristed needle driver. *Yonsei Med J.* 2016;57(4):1029–1033. Available from: https://doi.org/10.3349/ymj.2016.57.4.1029.
18. Matanes E, Lauterbach R, Mustafa-Mikhail S, Amit A, Wiener Z, Lowenstein L. Single port robotic assisted sacrocolpopexy: our experience with the first 25 cases. *Female Pelvic Med Reconstr Surg.* 2017;23(3):e14–e18. Available from: https://doi.org/10.1097/SPV.0000000000000397.
19. Ganesan V, Goueli R, Rodriguez D, Hess D, Carmel M. Single-port robotic-assisted laparoscopic sacrocolpopexy with magnetic retraction: first experience using the SP da Vinci platform. *J Robot Surg.* 2020;14(5):753–758. Available from: https://doi.org/10.1007/s11701-020-01050-1.
20. Dy GW, Jun MS, Blasdel G, Bluebond-Langner R, Zhao LC. Outcomes of gender affirming peritoneal flap vaginoplasty using the da Vinci single port vs Xi robotic systems. *Eur Urol.* 2021;79(5):676–683. Available from: https://doi.org/10.1016/j.eururo.2020.06.040.

CHAPTER

Future perspectives: natural orifice transluminal endoscopic surgery

11

Alireza Aminsharifi[1], Riccardo Bertolo[2], Christian Huge[1] and Wesley M. White[3]

[1]*Department of Urology, Hershey Medical Center, Pennsylvania State University, Hershey, PA, United States*

[2]*Department of Urology, San Carlo di Nancy Hospital, Rome, Italy*

[3]*Division of Urologic Surgery, The University of Tennessee Medical Center, Knoxville, TN, United States*

11.1 Introduction

As minimally invasive surgery (MIS) increases in utility given the shorter hospital stays and the potentially lower surgical morbidity, surgeons will continue to strive for more efficacious techniques. This has been evidenced by the rapid shift away from open surgery and to laparoscopic and now robotic-assisted surgeries. The appeal towards MIS, particularly with the advent of robotics, has been the idea that improved visibility and the more precise surgical techniques will allow for (1) decreased postoperative morbidity, (2) shorter hospital stay and quicker recovery, and (3) potentially fewer complications.[1]

In recent years, the adoption of newer technologies such as single-port (SP) robotic surgery performed by using a dedicated platform has shown promise in achieving these goals.[2,3] Early evidence suggests that with the use of single-site surgery, postoperative morbidity in terms of the need for narcotics, hospital stay, and overall recovery would be significantly decreased.[4,5] Concomitant with these advancements, the concept of scarless surgery has been always under investigation during the past two decades.

One such novel technique that has been viewed as predominantly experimental since its conception is natural orifice transluminal endoscopic surgery (NOTES). This technique is based on the principle of performing a surgery through a naturally hollow viscus, thereby essentially eliminating external scars for the patient, and potentially decreasing the morbidity associated with skin incisions.[6] Ultimately, the goals are the same, namely the improved cosmesis, paired to less postoperative pain and shorter recovery time.[1] Examples of entry points for NOTES procedures include transgastric, transvaginal (TV), transrectal, and transvesical.

In this chapter, we will review the background of NOTES procedures, and discuss some landmark experiments and studies regarding the application of NOTES

Single-Port Robotic Surgery in Urology. DOI: https://doi.org/10.1016/B978-0-323-91906-7.00007-7

procedures in urology, specifically focusing on the envelopes pushed in the field after the advent of the SP dedicated surgical platforms. Challenges and future directions will also be addressed.

11.2 Background—transition from laparoscopy to hybrid and pure NOTES

The first NOTES procedures were reported in the 1970s within the field of gynecology, where TV culdoscopic sterilization was described.[7,8] However, the term NOTES was not coined until the early 2000s and gained a larger audience after the Natural Orifice Surgery Consortium for Assessment and Research was established, and their "White Paper" was published highlighting critical areas that required further research and analysis.[9]

As various fields continued to advance NOTES techniques, it was quickly realized that standard laparoscopic instruments sometimes lacked the required flexibility and degrees of freedom to perform an entire operation safely and effectively.[6,10] Therefore some surgeries would require at least one laparoscopic assistant port or a small laparotomy for assistance or specimen extraction (the so called "hand-assisted NOTES"). This required a distinction to be made between pure-NOTES (done solely through the incision of the viscus) versus the hybrid-NOTES (that required assistance via laparoscopy or laparotomy).[10] In the pursuit of completing predominately pure NOTES procedures, many technological advances have been made to eliminate the ergonomic issues associated with standard laparoscopic instruments. For example, in 2006 Zeltser et al. developed a SP magnetic anchoring and guidance system that allowed them to successfully perform a standard laparoscopic nephrectomy with minimal difficulty in manipulating instruments compared to that encountered when using a standard equipment.[11] Additionally, simple changes such as different surgical positioning has been shown to be effective.[12] When implementing the use of robotics, it was found that pure NOTES procedures were highly impractical given the extensive extracorporeal instrument clashing in the confined surgical space in the lithotomy position.[13] Laydner et al. found that repositioning to the prone-jack-knife position was able to mitigate the instrument clashing for performing NOTES nephrectomy (Fig. 11.1).[12]

11.2.1 NOTES in urology

In the field of urology, the concept of NOTES was first described by Breda et al. in 1993 with the vaginal extraction of an intact kidney after laparoscopic nephrectomy.[13] It was not until 2002 that NOTES procedures were performed in live animal models by Gettman et al.[14] In this small series of six TV laparoscopic nephrectomies in female farm pigs, five were hybrid NOTES and one was pure NOTES. All hybrid NOTES procedures were completed with only a single 5 mm

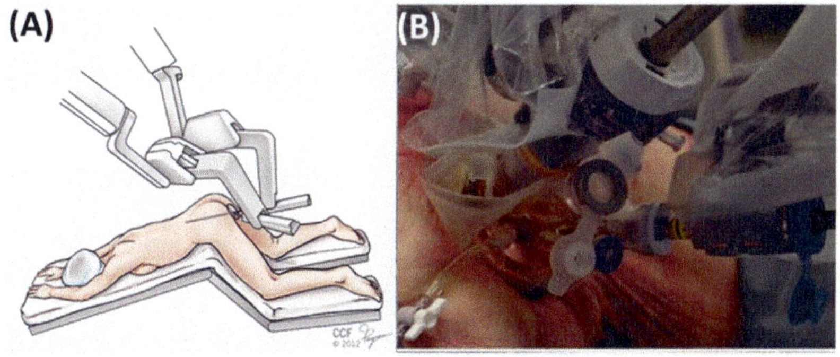

FIGURE 11.1

(A) The prone-jack-knife position; (B) The QuadPort + allocating the robotic instruments of a Si system.

Adapted from https://doi.org/10.1016/j.urology.2012.11.085 UROLOGY 81: 1238, 2013. Copyright 2013 Elsevier Inc.

trans-abdominal trocar with mean operative time of 210 minutes. For the pure NOTES procedures, the surgery took 360 minutes and the animal survived with no postoperative or intraoperative complications. These results were quite promising and emboldened surgeons to continue translational work.

Aminsharifi et al. expanded upon animal trials in TV NOTES nephrectomy using 10 female dogs for both right and left nephrectomies. Like Gettman et al., these surgeries showed great promise with minimal intraoperative blood loss (mean change in pre- and post-operative hemoglobin 0.31) and operative times ranging from only 75−135 minutes.[14,15] Abdominal re-exploration 1 month postoperatively showed no significant intra-abdominal adhesions and complete healing of the colpotomy incision with this technique.

Continued experimental surgeries in both animal and cadaver models expanded the field with surgeries including NOTES radical prostatectomy and combined transgastric and transurethral NOTES partial cystectomy.[16,17] In 2008, Branco et al. reported the first TV hybrid NOTES to be performed in a live patient (Fig. 11.2).[18]

In 2009, Kaouk et al. were the first to perform a TV pure NOTES nephrectomy.[19] Currently, many groups have reported successful pure and hybrid NOTES surgeries in clinical practice with safe and effective outcomes.[20] A cumulative retrospective analysis of NOTES and natural orifice specimen extractions (NOSE) was reported by Zou et al. including 35 animal and 305 clinical surgeries.[20] Within this study, both pure and hybrid TV NOTES were included and procedure variety included nephrectomy, nephroureterectomy, partial nephrectomy, cystectomy, and adrenalectomy. These surgeries were undertaken for both malignant and benign conditions, with all oncologic surgeries having tumor size <7 cm (cT2a). In total, 12 TV-NOSE, 260 hybrid TV-NOTES, and 25 pure TV-NOTES were

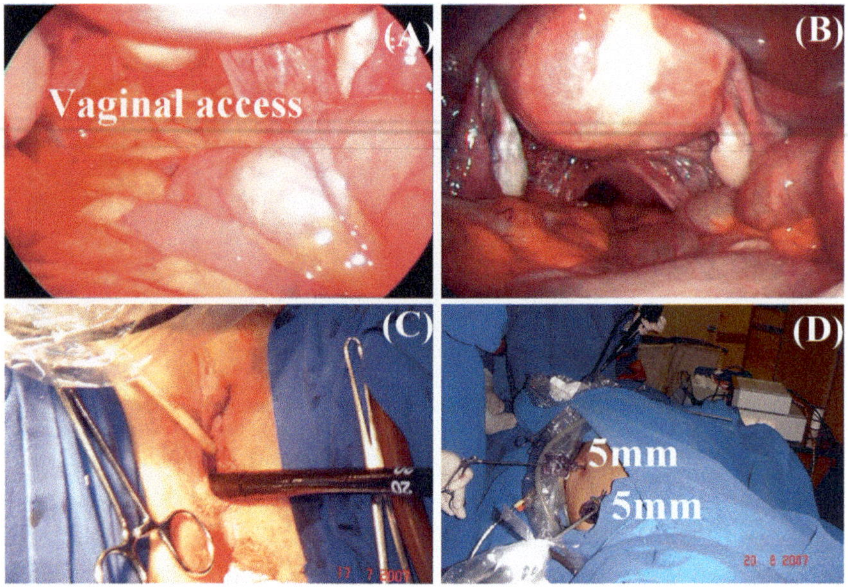

FIGURE 11.2

(A) Index finger dissection of the cul-de-sac and entrance in the abdominal cavity. (B) Laparoscopic visualization of the pelvic structures. (C) Placement of the endoscope through the vaginal cavity. (D) External view of the surgery.

From https://doi.org/10.1016/j.eururo.2007.10.053. Copyright 2008 European Association of Urology. Published by Elsevier B.V. All rights reserved.

collected with major complications occurring in 7.2% of all cases. The causes of complications related to the learning curve, the presence of dense pelvic adhesions, or rectal injury ($n = 1$) during blind placement of the vaginal port. The procedures were tolerated well with minimal pain and similar preoperative and postoperative female sexual function index scores. Four patients were able to conceive naturally within 1 year of surgery. No complications with delivery secondary to previous vaginal access were observed. It must be emphasized that given the technical complexity and skillset required to achieve the aforementioned successes, these surgeries are still considered experimental and further research and analysis of clinical trials is required before implementing into standard practice.

11.2.1.1 NOTES nephrectomy

11.2.1.1.1 Trans-vaginal access for NOTES nephrectomy

Nephrectomy is the most common urological NOTES. Branco et al. performed the first clinical TV hybrid NOTES nephrectomy in 2008 for a 23-year-old female with a nonfunctional right kidney secondary to chronic urinary tract infections.[18] Their technique utilized two 5 mm abdominal trocars in addition to the TV ports.

The operative time was 170 minutes and the patient was able to be discharged on postoperative day one and return to her daily routines on day three. This was a landmark report for the field of NOTES in urology and showed that it could be safely performed in the clinical setting.

The next advancement came in 2009 with successful pure TV NOTES nephrectomy in the clinical setting by Kaouk et al. (Figs. 11.3 and 11.4).[19]

Nephrectomy was done for a 58-year-old female with a history of right vesicoureteral reflux, recurrent urinary tract infections, and a resultant poorly functioning right kidney. The entirety of the procedure was completed trans-vaginally with operative time of 420 minutes and an estimated blood loss of only 50 cc. Importantly, the authors noted difficulties faced during this case were the close proximity of instrument ports resulting in extracorporeal clashing and poor tissue handling. Additionally, despite being able to complete the surgery completely through a vaginal access, the authors recommended that, atleast in early experience, the TV access should be done under direct visualization via a standard laparoscopic port camera to reduce the risk of visceral injury with a blind access through vagina.

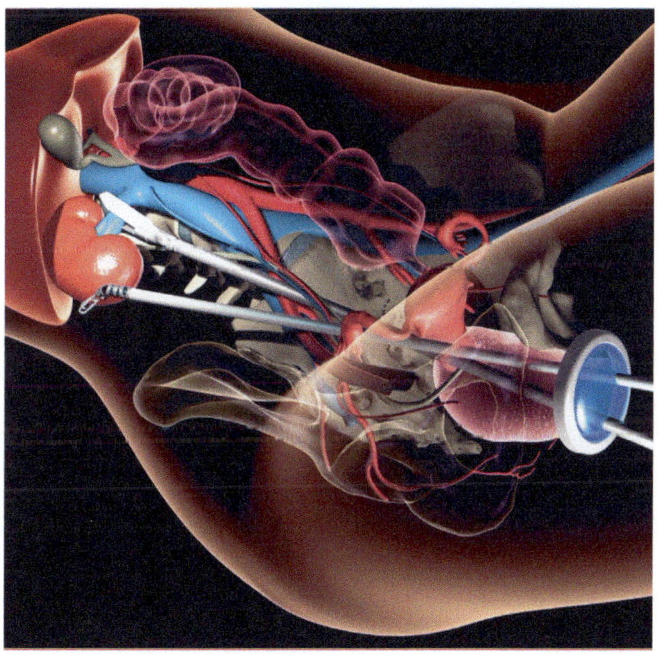

FIGURE 11.3

Cartoon showing the concept of pure NOTES nephrectomy.

*From https://doi.org/10.1016/j.eururo.2009.10.027. *

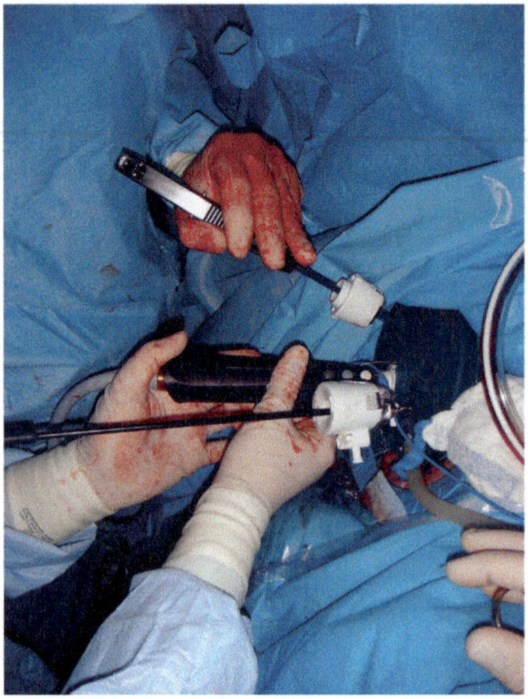

FIGURE 11.4

Kaouk and colleagues performing the first pure NOTES nephrectomy.

From https://doi.org/10.1016/j.eururo.2009.10.027. Copyright 2009 European Association of Urology. Published by Elsevier B.V. All rights reserved.

As technique and experience continued to improve, another landmark case of hybrid TV NOTES nephrectomy was presented in 2012 when Kaouk et al. performed the first clinical TV hybrid NOTES transplant donor nephrectomy.[21] Their technique was quite novel as they performed the surgery robotically, using the robotic arms for both the umbilical and the vaginal incisions (Fig. 11.5).

This configuration allowed them to resolve the constraints of extra-corporeal clashing seen in TV pure NOTES. They were able to complete the procedure without the need for conversion to standard laparoscopy with an operative time of 240 minutes, a warm ischemia time of 5.8 minutes and an EBL of 75 cc. The patient tolerated the procedure well and was discharged after 48 hours of observation and required no intravenous pain medications. At follow-up of 4 weeks postoperatively, there were no complaints of vaginal pain or drainage, a well-healed vaginal incision, and patients were allowed resumption of regular sexual activity at that time. Although feasible, the author noted that further trials would have been necessary to evaluate safety and efficacy of this approach compared to standard laparoscopy.

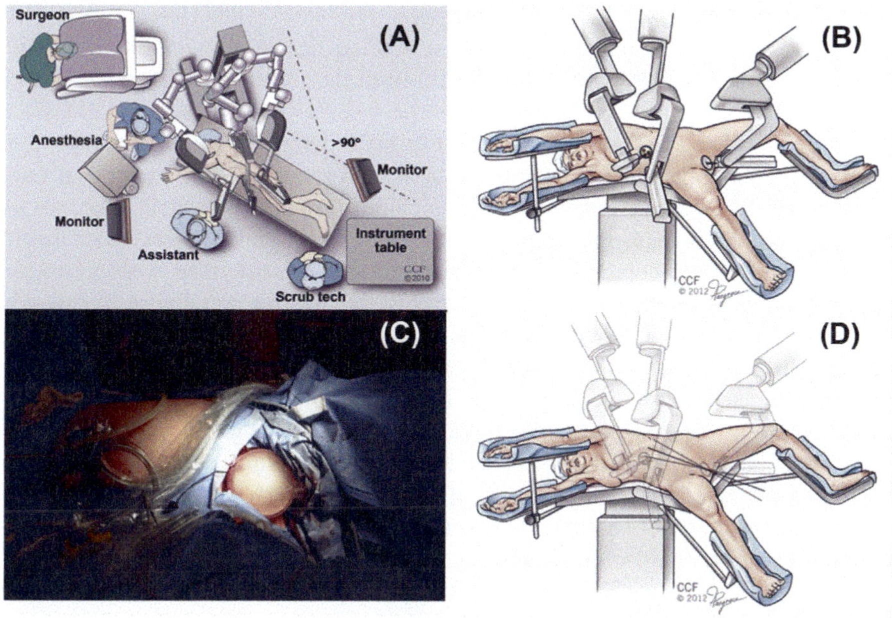

FIGURE 11.5

(A) Operative room setup. (B) Patient positioning. (C) SILS port site in umbilical incision and GelPoint port site in vagina. (D) da Vinci-Si robot docked from behind patient's back. Right robot arm docked to vaginal port, left robot arm docked to transumbilical port, and scope inserted through 12-mm transumbilical SILS port. Fourth robot arm parked away from surgical field and not used with range of motion of robotic arms.

From https://doi.org/10.1016/j.urology.2012.08.061. Copyright 2012 Elsevier Inc.

Kaouk et al. actually paved the way for robotic NOTES. They were the first engrafting a robotic platform to the NOTES approach, trying to bridge the technology gap. Their hybrid approach using a robotic platform facilitated the surmounting of the technical challenges noted in both LESS (the instrument clashing) and NOTES. Nevertheless, the approach did not know further widespread diffusion.

Until this point, most renal surgeries performed via NOTES approach, whether hybrid or pure, were for benign indications. However, in 2014 Georgiopoulos et al. published their results on 38 hybrid TV-NOTES nephrectomies for tumor (23) and benign (15) indications.[22] Their average demographics included average tumor size of 6 cm (largest of 7.5 cm) for renal cell carcinoma. Average operative time was 114.1 minutes with EBLs ranging from 50–150 cc. Intraoperative bladder injuries occurred in two patients that were closed primarily and two patients required transfusions postoperative. Notably, no positive surgical margins were noted after pathology reports.

11.2.1.1.2 Trans-rectal and trans-gastric access for NOTES nephrectomy

Trans-vaginal NOTES has been the most studied access point thus far given a sheer ease of access. Obviously, this would limit the use of NOTES procedures in male population.

Success with gastro-intestinal access has been demonstrated in several animal and cadaver model trials. In 2011, Bazzi et al. described the first attempts at trans-rectal NOTES via a hybrid transrectal approach in a porcine model.[23] Access was obtained through a 2-cm incision above the dentate line under direct visualization via umbilical laparoscopic port. A total of three nephrectomies were performed with median operative time of 180 minutes (30 minutes for rectal access) and average blood loss less than 50 cc. The rectal incision was closed primarily without significant difficulty. Necropsy was performed 1 week postoperatively that noted no intra-abdominal injuries or signs of peritonitis. Similar results were seen in Park et al., where 15 female pigs were randomly divided into groups to undergo TV, transrectal, and standard laparoscopic nephrectomy.[24] Mean operative times were significantly longer for TV and transrectal approaches (61 and 84 minutes, respectively) compared to conventional laparoscopy (24 minutes), but no evidence of visceral injury or peritonitis was noted on postmortem exams in all three groups.

Data on trans-gastric NOTES in urology is quite limited compared to trans-vaginal and even trans-rectal surgeries. In 2007, Lima et al. published their experience in a nonsurvival study of six trans-gastric and trans-vesical nephrectomies in porcine models.[25] Median operative time was 120 minutes but notably decreased with experience per author report. Complications included hemorrhage during the dissection of the renal vasculature. Importantly, the specimens were not extracted from the pigs and the gastrotomy was not closed due to the lack of appropriate instrumentation. This topic was revisited in 2010 when Boylu et al. performed a pure transgastric NOTES partial nephrectomy in a single porcine model with specimen extraction and closure of the gastrotomy.[26] They performed left upper pole nephrectomy with an operative time of 240 minutes and complete hemostasis was achieved with estimated blood loss of 240 cc. This was confirmed with direct visualization of the surgical site 4 hours postoperatively.

11.2.1.2 NOTES prostatectomy

11.2.1.2.1 Trans-urethral approach for NOTES Prostatectomy

To expand the application of NOTES procedures in urology, Humphreys et al. tried to develop NOTES radical prostatectomy with the urethra as the natural access point. They applied the concept and principles of the Holmium laser enucleation of the prostate (HoLEP) for performing a radical prostatectomy.[16] To prove this concept, this group performed NOTES radical prostatectomy in four male cadavers using a surgical set-up similar to that of HoLEP. Intraoperatively, they were able to preserve bilateral neurovascular bundles, the dorsal venous complex, the bladder neck, and the external sphincter.[16] The vesicourethral anastomosis was completed using a laparoscopic suture device and knot pusher system. After excision of the prostate, the prostate was morcellated within the bladder. Negative margins were found at the NVB, bladder neck, and urethra. Despite the technical feasibility, questions about the possibility of cancer seeding in the bladder from the morcellation of cancerous prostatic tissue as well as the inability to perform appropriate lymph-nodes dissection when indicated remained.

In 2011 and 2012, two separate studies investigated the first clinical applications of NOTES radical prostatectomy. Humphreys et al. were the first who used their original technique on cadaver models.[27] They extracted the en-bloc specimen via a small suprapubic cystotomy. This surgery was completed in two separate patients, one with Gleason $3 + 3 = 6$, Stage pT2a Nx Mx and the other with Gleason $3 + 4 = 7$, Stage pT2c Nx Mx. Negative margins did result after pathologic review. Patient one was able to have his catheter removed after 7–10 days. A similar concept was presented by Nagele et al. who reported a single NOTES radical prostatectomy (Gleason $4 + 3 = 7$, Stage pT2c Nx Mx) using a thulium laser.[28] They performed a sutureless vesicourethral anastomosis via traction with a Foley catheter and placement of a supra-pubic tube for improved drainage.

11.2.1.3 Trans-rectal approach NOTES prostatectomy

In 2015, Akca et al. reported successful trans-rectal NOTES radical prostatectomy in a cadaver model using SP robotics.[29] Authors noted ease of exposure of the posterior prostate and seminal vesicles as well as extraction of an intact specimen through the trans-rectal port site. This is in stark contrast to the transurethral approach, which necessitates cystotomy to retrieve the prostate in the setting of oncologic surgery.[27] There were no reported injuries to vascular, visceral, or neural structures during this procedure and the primary rectal incision was able to be closed securely.

11.3 NOTES in urology: challenges, limitations, and future directions

11.3.1 Inadequate instrumentation

At the present time, there are very limited dedicated instruments and platforms for NOTES procedures. The use of standard minimally invasive instruments for NOTES can be challenging due to their limited degrees of freedom and maneuverability for surgery performed through natural orifices. For instance, Laydner et al. addressed how the poor flexibility, in addition to the close proximity of the ports resulted in extra-corporeal clashing in the dorsal lithotomy position. These constrains have thus far made TV surgery, even performed by using robotic platforms, highly impractical (Fig. 11.6).[12]

To manage this problem, they performed two cadaver nephrectomies with one in dorsal lithotomy and another in prone-jack-knife position. While they were unable to complete the procedure in dorsal lithotomy due to poor bowel retraction

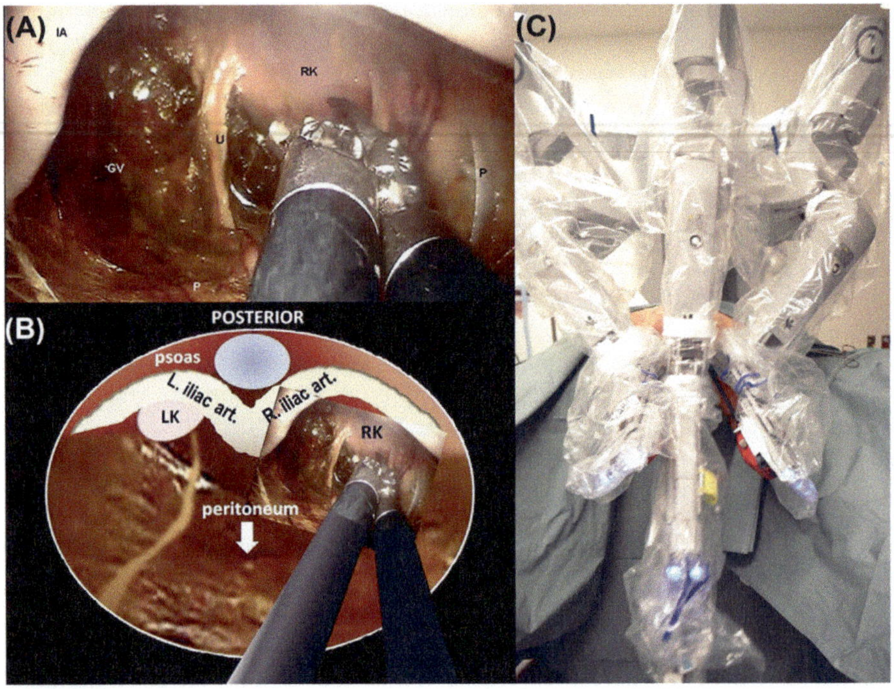

FIGURE 11.6

(A) Intraoperative view: iliac artery; right kidney; ureter; gonadal vein; peritoneum. (B) Panel A inserted into the context of a schematic representation of the whole retroperitoneal cavity; left kidney. (C) External view.

From https://doi.org/10.1016/j.urology.2012.11.083.

and constant clashing with the patient's legs, they were able to successfully perform a robotic NOTES nephrectomy through the vagina using a retroperitoneal approach in the prone jack-knife position (see Fig. 11.1). This was certainly a step forward in advancing the incorporation of robotic technology into NOTES. On the other hand, it witnessed that further research was still required for improvement in instrumentation and technique.

As such, the approach was born out of necessity. The prone position of the patient firstly left room for the bulky robotic arms, thus potentially limiting the clashing of the instruments with the patient's legs. Secondly, the retroperitoneal access obviated the need for bowel retraction.

On the other hand, many aspects of this unfamiliar anatomic approach were disorienting. Indeed, the authors were only able to complete one of the three attempted cases.

It will be eventually in the hands of the growing advances in SP robotic surgery to bring renewed indications for NOTES procedures.

11.3.2 Inadequate data for standardization

Although NOTES held promise as evidenced by the aforementioned studies, the sheer volume of cases for comparison against the gold standard is lacking. Larger studies such as that performed by Zou et al. may answer questions regarding the efficacy but leave many more questions remaining.[20] The entirety of research to this point has been purely experiential and lacked control groups. Comments regarding safety and feasibility are therefore subjective and may not be translatable. Notably, data specifically regarding robot-assisted NOTES are even more anecdotal than those about laparoscopic NOTES.

11.3.3 Oncologic safety

While most of the NOTES procedures have been performed for benign disease conditions, there are a few reports that evaluate the safety and efficacy of NOTES procedures for urological malignancies. Despite the small sample sizes, the preliminary data has shown promise. For renal carcinoma, negative margins for both partial and radical nephrectomies have been able to be achieved with TV NOTES, albeit in series with small sample size.[20,22] In the realm of prostate cancer, successful radical prostatectomies have been performed via either a transurethral or a transrectal approach, but likewise within the setting of small series.[27–29] However, lymph-nodes dissection is not feasible with current NOTES procedures for radical prostatectomy. Therefore this procedure can be offered to highly selected patients only. Long-term follow-up data as well as comparative trials are required to better assess the oncological outcomes of NOTES procedure compared to those of the "standard" techniques.

11.4 Future directions

Although beyond the scopes of urologic surgery, a special mention goes to general surgery. NOTES has been extensively adopted for dealing with the challenge of rectal cancer surgery. Robotic transanal surgery ushered in a new age of three-dimensional visualization, wristed movement, and ergonomics. However, due to the troubles of operating within the bony confines of the narrow pelvis, the multiarm surgical platform presented new technological challenges, including bulky arms and difficulty sustaining an airtight seal to maintain insufflation and visualization within the rectum.

Despite these challenges, the Xi robot was utilized to perform transanal total mesorectal excision with complete or near complete mesorectal excisions for complex fistula repair and invasive adenocarcinoma of the distal rectum.[30] This demonstration of robotic transanal surgery not only confirmed its utility in the surgical excision of rectal cancer, but also highlighted that the Si and Xi platforms were best designed for other approaches, namely the transabdominal, multitrocar approach.[31]

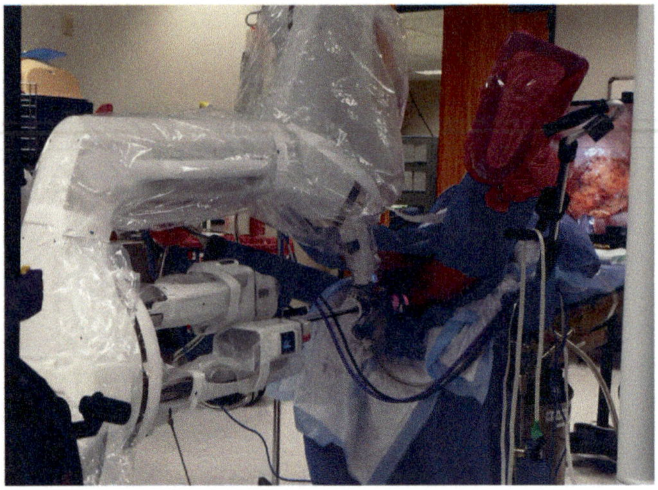

FIGURE 11.7

Robotic transanal surgery performed on a cadaver using the da Vinci SP surgical system (Intuitive, Sunnyvale, California, United States) through the applied GelPOINT path transanal access platform. Insufflation was created using a valveless trocar system (ConMed, Inc). The cadaver was placed in the dorsal lithotomy position.

From https://doi.org/10.1007/s10151-017-1655-3. Copyright Springer International Publishing AG 2017.

In 2018, the da Vinci SP robot (Intuitive Surgical, Sunnyvale, California, United States) was developed to accommodate narrow orifice procedures through its single-arm, SP system in which three 6-mm, multijointed wristed instruments and a fully wristed 3D HD camera can be, for example, inserted transanally via the dedicated 25-mm cannula.

In preclinical models, this system was utilized to perform transanal local excisions with negative margins and no fragmentation, as well as endoluminal suture closure of the defects, providing encouraging evidence for its use in the field of transanal surgery (Fig. 11.7).[32]

In clinical practice, the use of this platform in the field of colorectal and general surgery has already been reported to perform left colectomy, right colectomy, and transanal minimally invasive procedures.[33–35]

To date, there has been no report detailing the feasibility of the adoption of the da Vinci SP platform for performing transanal total mesorectal excisions until the experience published by Marks et al.[36] The authors aimed to assess both the safety and the feasibility of the use of the da Vinci SP surgical device in performing SP robotic transanal total mesorectal excision.

They published the data of two patients. Both the operations were completed successfully without complications or conversion. In both the cases the total mesorectal excision was completed transanally using the SP robot. In the first patient, the abdominal portion was completed through an abdominal single-incision, whilst in the second patient the operation was entirely performed transanally as a pure NOTES procedure (Figs. 11.8 and 11.9).

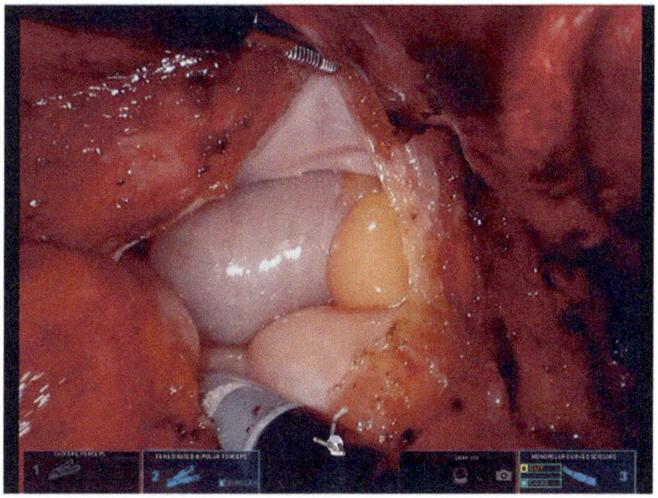

FIGURE 11.8

Single-port robotic transanal total mesorectal excision entrance into the peritoneal cavity.

From https://doi.org/10.1007/s10151-021-02449-0. Copyright Springer Nature Switzerland AG 2021.

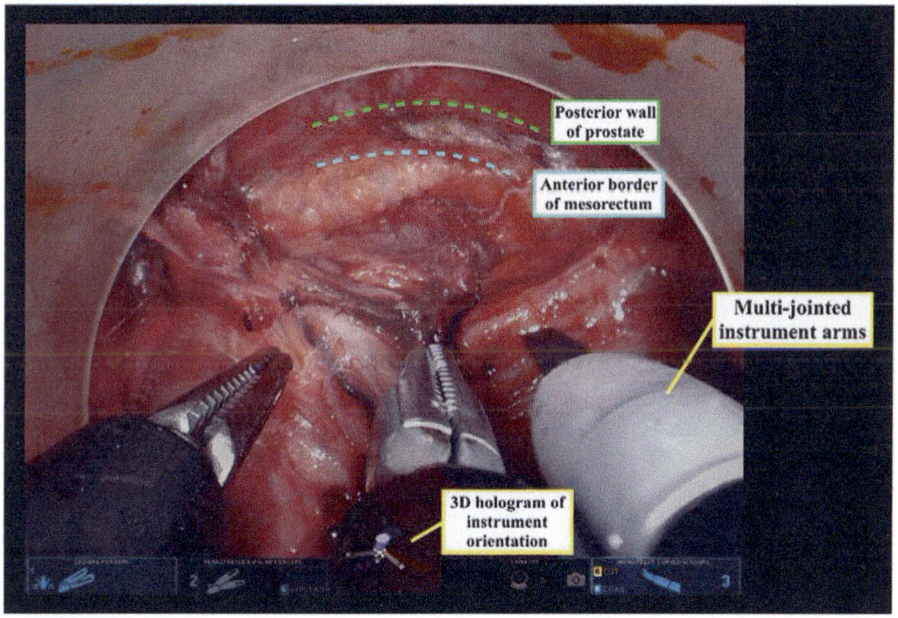

FIGURE 11.9

Robotic console view during single-port robot-assisted NOTES transanal total mesorectal excision performed by using the da Vinci SP platform.

From https://doi.org/10.1007/s10151-021-02449-0. Copyright Springer Nature Switzerland AG 2021.

Based on their preliminary experience, the authors concluded that SP robot-assisted NOTES transanal total mesorectal excision was feasible and safe by using the da Vinci SP platform, with excellent optics and dexterity to work in such a limited surgical space.

11.5 Conclusions

Any new medical technology or surgical procedure demands a stringent and objective evaluation of its merit, pragmatism, and reproducibility. These metrics must then be weighed against the potential benefit advancing that may be offered to the patient. As demonstrated in this chapter, NOTES is an intriguing surgical approach which might offer patients the potential for optimized cosmesis, accelerated recovery, and decreased pain.

Various minimally invasive surgical approaches have been utilized using laparoscopic instruments and techniques. These have been associated with difficulty in exposure, reach and ergonomics, leading to a high conversion rate and intraoperative troubles in general.

Robotic surgery was applied to NOTES in hopes of providing benefits of improved dexterity and operative precision with a stable and ergonomic platform, but multiarms robotic platforms failed in expanding the indications of NOTES approach.

Indeed, while other robotic systems have been developed to address issues regarding configuration of multiple robotic arms, the recently introduced da Vinci SP robotic surgical system is the first of its kind specifically designed for single incision and endoluminal surgery.

The momentum is gathering albeit slowly. As such, the few data from the historical series about laparoscopic NOTES in urology have not yet given renewed impetus to NOTES.

Ultimately, many questions remain unanswered regarding optimal patient selection, safety of entry and closure when applied across a large population, and cost effectiveness.

Disclosures

None.

References

1. Swain P. A justification for NOTES—natural orifice translumenal endosurgery. *Gastrointest Endosc*. 2007;65(3):514–516.

2. Kaouk J, Aminsharifi A, Sawczyn G, et al. Single-port robotic urological surgery using purpose-built single-port surgical system: single-institutional experience with the first 100 cases. *Urology*. 2020;140:77–84.
3. Aminsharifi A, Sawczyn G, Wilson CA, Garisto J, Kaouk J. Technical advancements in robotic prostatectomy: single-port extraperitoneal robotic-assisted radical prostatectomy and single-port transperineal robotic-assisted radical prostatectomy. *Transl Androl Urol*. 2020;9(2):848.
4. Wilson CA, Aminsharifi A, Sawczyn G, et al. Outpatient extraperitoneal single-port robotic radical prostatectomy. *Urology*. 2020;144:142–146.
5. Sawczyn G, Lenfant L, Aminsharifi A, Kim S, Kaouk J. Predictive factors for opioid-free management after robotic radical prostatectomy: the value of the SP® Robotic Platform. *Minerva Urol Nephrol*. 2020;73:591–599.
6. Kobiela J, Stefaniak T, Mackowiak M, Lachinski AJ, Sledzinski Z. Third generation surgery. Vain hopes or the reality of tomorrow? *Langenbeck's Arch Surg*. 2008;393(3):405–411.
7. McCann MF, Cole LP. Risks and benefits of culdoscopic female sterilization. *Int J Gynecol Obstet*. 1978;16(3):242–247.
8. Denniston GC. Sterilization by culdoscopy. *J Reprod Med*. 1973;10(4):175–176.
9. Clark MP, Qayed ES, Kooby DA, Maithel SK, Willingham FF. Natural orifice translumenal endoscopic surgery in humans: a review. *Minim Invasive Surg*. 2012;2012:189296.
10. Tyson MD, Humphreys MR. Urological applications of natural orifice transluminal endoscopic surgery. *Nat Rev Urol*. 2014;11(6):324–332.
11. Zeltser IS, Bergs R, Fernandez R, Baker L, Eberhart R, Cadeddu JA. Single trocar laparoscopic nephrectomy using magnetic anchoring and guidance system in the porcine model. *J Urol*. 2007;178(1):288–291.
12. Laydner H, Autorino R, Isac W, et al. Robotic retroperitoneal transvaginal natural orifice translumenal endoscopic surgery (NOTES) nephrectomy: feasibility study in a cadaver model. *Urology*. 2013;81(6):1232–1238.
13. Breda G, Silvestre P, Giunta A, Xausa D, Tamai A, Gherardi L. Laparoscopic nephrectomy with vaginal delivery of the intact kidney. *Eur Urol*. 1993;24:116–117
14. Gettman MT, Lotan Y, Napper CA, Cadeddu JA. Transvaginal laparoscopic nephrectomy: development and feasibility in the porcine model. *Urology*. 2002;59(3):446–450.
15. Aminsharifi A, Taddayun A, Shakeri S, Hashemi M, Abdi M. Hybrid natural orifice transluminal endoscopic surgery for nephrectomy with standard laparoscopic instruments: experience in a canine model. *J Endourol*. 2009;23(12):1985–1989.
16. Humphreys MR, Krambeck AE, Andrews PE, Castle EP, Lingeman JE. Natural orifice translumenal endoscopic surgical radical prostatectomy: proof of concept. *J Endourol*. 2009;23(4):669–675.
17. Sawyer MD, Cherullo EE, Elmunzer BJ, Schomisch S, Ponsky LE. Pure natural orifice translumenal endoscopic surgery partial cystectomy: intravesical transurethral and extravesical transgastric techniques in a porcine model. *Urology*. 2009;74(5):1049–1053.
18. Branco AW, Branco Filho AJ, Kondo W, et al. Hybrid transvaginal nephrectomy. *Eur Urol*. 2008;53(6):1290–1294.
19. Kaouk JH, Haber GP, Goel RK, et al. Pure natural orifice translumenal endoscopic surgery (NOTES) transvaginal nephrectomy. *Eur Urol*. 2010,57(4):723–726.
20. Zou X, Zhang G, Xie T, et al. Natural orifice transluminal endoscopic surgery in urology: the Chinese experience. *Asian J Urol*. 2020;7(1):1–9.

21. Kaouk JH, Khalifeh A, Laydner H, et al. Transvaginal hybrid natural orifice transluminal surgery robotic donor nephrectomy: first clinical application. *Urology*. 2012;80 (6):1171–1175.
22. Georgiopoulos I, Kallidonis P, Kyriazis I, et al. Hybrid transvaginal nephrectomy: development of our technique. *Urology*. 2014;84(1):99–105.
23. Bazzi WM, Wagner O, Stroup SP, et al. Transrectal hybrid natural orifice transluminal endoscopic surgery (NOTES) nephrectomy in a porcine model. *Urology*. 2011;77 (3):518–523.
24. Park YH, Kim KT, Bae JB, Kim HH. Transvaginal and transrectal natural orifice translumenal endoscopic surgery nephrectomy in a porcine survival model: comparison with conventional laparoscopic nephrectomy. *J Endourol*. 2015;29(3):351–356.
25. Lima E, Rolanda C, Pêgo JM, et al. Third-generation nephrectomy by natural orifice transluminal endoscopic surgery. *J Urol*. 2007;178(6):2648–2654.
26. Boylu U, Oommen M, Joshi V, Thomas R, Lee BR. Natural orifice translumenal endoscopic surgery (NOTES) partial nephrectomy in a porcine model. *Surg Endosc*. 2010;24(2):485–489.
27. Humphreys MR, Sauer JS, Ryan AR, et al. Natural orifice transluminal endoscopic radical prostatectomy: initial perioperative and pathologic results. *Urology*. 2011;78 (6):1211–1217.
28. Nagele U, Anastasiadis AG, Walcher U, Nicklas AP, Merseburger AS, Herrmann TR. Natural orifice (NOTES) transurethral sutureless radical prostatectomy with thulium laser support: first patient report. *World J Urol*. 2012;30(5):625–631.
29. Akça O, Zargar H, Autorino R, et al. The transrectal single port laparoscopic radical prostatectomy in a cadaver model. *Turkish J Urol*. 2015;41(2):78.
30. Atallah S, Nassif G, Polavarapu H, et al. Robotic-assisted transanal surgery for total mesorectal excision (RATS-TME): a description of a novel surgical approach with video demonstration. *Tech Coloproctol*. 2013;17(4):441–447. Available from: https://doi.org/10.1007/s10151-013-1039-2.
31. Atallah S, Hodges A, Larach SW. Direct target NOTES: prospective applications for next generation robotic platforms. *Tech Coloproctol*. 2018;22(5):363–371. Available from: https://doi.org/10.1007/s10151-018-1788-z.
32. Marks J, Ng S, Mak T. Robotic transanal surgery (RTAS) with utilization of a next-generation single-port system: a cadaveric feasibility study. *Tech Coloproctol*. 2017; 21(7):541–545. Available from: https://doi.org/10.1007/s10151-017-1655-3.
33. Marks JH, Kunkel E, Salem J, Martin C, Schoonyoung HP, Agarwal S. rSILS: initial clinical experience with single-port robotic (SPr) right colectomy. *Tech Coloproctol*. 2020;24(8):817–822. Available from: https://doi.org/10.1007/s10151-020-02218-5.
34. Marks JH, Salem JF, Anderson BK, Josse JM, Schoonyoung HP. Single-port robotic left colectomy: first clinical experience using the SP robot (rSILS). *Tech Coloproctol*. 2020;24(1):57–63. Available from: https://doi.org/10.1007/s10151-019-02129-0.
35. Marks JH, Kunkel E, Salem JF, Martin C, Anderson B, Agarwal S. First clinical experience with single-port robotic transanal minimally invasive surgery (SP rTAMIS) for benign rectal neo- plasms. *Tech Coloproctol*. 2021;25(1):117–124. Available from: https://doi.org/10.1007/s10151-020-02358-8.
36. Marks JH, Salem JF, Adams P, et al. SP rTaTME: initial clinical experience with single-port robotic transanal total mesorectal excision (SP rTaTME). *Tech Coloproctol*. 2021;25(6):721–726. Available from: https://doi.org/10.1007/s10151-021-02449-0. Epub 2021 Apr 21.

Index

Note: Page numbers followed by "*f*" and "*t*" refer to figures and tables, respectively.

T